"Dramatic and fair-minded. Dr. Hazen writes with wonderful clarity about science.... Effortlessly teaches as it zips along."

Christopher Lehmann-Haupt
The New York Times

"Fascinating...Too few scientists are willing to humble themselves, and their writing, to translate such wonder for the layman."

Gayle Golden
The Dallas Morning News

"Shows a side of science few would want to believe exists: the intellectual sparring, betrayals, deceits and the almost ceaseless competition among some to become scientific stars."

Jeane Malone
The Decatur Daily

"He reveals the paranoia, his own misgivings, the wealth of rumor, the patent fights, the fear of being wrong.... An exhilirating account of one of the most exciting technological developments of the century."

The Kirkus Reviews

"An intensely personal, nuts-and-bolts account ...Class-A popular science writing."

Booklist

The Breakthrough

The Race for the Superconductor

Robert M. Hazen

BALLANTINE BOOKS ● NEW YORK

ISBN 0-345-36145-8

This edition published by arrangement with Summit Books, a division of Simon & Schuster, Inc.

Manufactured in the United States of America

First Ballantine Books Edition: September 1989

To my friends and
colleagues of the
Geophysical Laboratory

Contents

Preface

Scientific research can be joyously unpredictable. Scientists begin their experiments with logical expectations and rational preconceptions, and sometimes their educated guesses turn out to be correct. But nature has a way of playing tricks on the researcher who tries to unlock her secrets; she will have her own way. Every scientist hopes for the good fortune to recognize one of nature's surprises and the good sense to make the most of it. That's how scientists become famous. Nature played such a trick on Paul Chu—and he made history.

For two decades Paul Chu pursued an intense and often frustrating program of research on superconductors—mysterious materials that become perfect conductors of electricity when chilled to unimaginably cold temperatures. Superconductors defy everyday experience. Almost all substances have some resistance to electrical currents. Even copper, which is used for most electrical wiring, subjects electricity to resistance, so much so that a significant fraction—in some cities as much as half—of all the electrical power generated is lost as heat in copper wires. Only a superconducting wire could prevent such losses, for a current that is started in a superconductor can flow forever.

Dozens of superconductor applications—some of which could change the fabric of society—have been on the drawing boards for years. Powerful and efficient lightweight magnets and motors, compact computers faster than any now in existence, magnetically levitated bullet trains able to travel hundreds of miles per hour, money-saving power transmission cables linked to safe, remote nuclear and solar generators, long-term energy storage systems, and a multitude of other devices are all possible in theory. But in practice, for everyday applications, it has been simply too difficult and too expensive to refrigerate any of the known superconducting devices to the requisite low "critical" temperatures.

Then in January 1986, Alex Müller and Georg Bednorz, two little-known researchers at IBM's Zurich research center, discovered zero resistivity at a record high temperature in a copper-bearing oxide material—a material that by conventional wisdom should not have been superconducting at all. At first, few physicists believed their results; there had already been so many false claims of high-temperature superconductivity that only a handful of researchers took the Zurich announcement seriously.

But Paul Chu believed. Inspired by published reports of the IBM advance, his team in Houston immediately went to work on the new compound. Laboring days and nights, weekends and holidays, they first tried to duplicate and then to surpass the Zurich result. During the winter of 1986–87 they performed experiment after experiment in the hopes of obtaining a slightly more efficient superconducting sample. What they found instead was an astonishing, totally new variety of superconductors that operated at much higher temperatures than most experts had thought possible—temperatures well within a range that allows practical applications. It was serendipity. In their systematic search for the best variant of the IBM materials they had discovered a previously unknown compound that would send the world's physics community into a frenzy of research and development.

As news of Chu's success spread—and it spread fast—thousands of scientists and engineers the world over dropped everything to concentrate on one of the most exciting scientific discoveries in decades. Unlike the transistor and the laser, which were invented before most applications had been conceived, high-temperature superconductors had been a dream of physicists for years. All that was missing was the mythical compound that would transform these dreams into reality. Chu's creation was the closest anyone had yet come to a practical superconductor, and so his discovery caused an unprecedented research stampede, a mad race to isolate, identify, and characterize the new material.

* * *

I was part of that race. *The Breakthrough* is an insider's view of the scientific process, spiced with the inevitable foibles of intense personalities and the machinations spawned of rivalry, as viewed from the Washington laboratory that I know best. It is not a historical overview of superconductor science and technology, nor even a comprehensive treatment of the new superconducting compound. Instead the book focuses primarily on the intensive research efforts.

The first part of the book recounts the dramatic discoveries made at the University of Houston and the University of Alabama laboratories during the winter of 1986–87. Chu knew that his synthesis of a remarkable and magical new superconductor could lead to billions of dollars of new technology within the decade. He also knew that his synthetic material was a mixture of at least two different chemical compounds, only one of which was the superconductor. His laboratory did not have the scientific hardware necessary to isolate and identify the specific superconducting material, but he needed that information in order to exploit his discovery.

The book's second section is a firsthand, day-by-day account of the activities of the Geophysical Laboratory team (of which I was a part), whose members raced to identify the chemical composition and crystal structure of Paul Chu's compound with its distinctive 1-2-3 ratio of chemical elements. The tale is a microcosm of the scientific process—a process that usually takes many months or years of exacting experimentation, theoretical analysis, and product development. But in this case experiments and theorizing, research and development, and even the first attempts at historical analysis were concentrated into a few short weeks of frantic labor. For many of us who shared in that hectic time of round-the-clock research, it was the most exciting experience of our professional lives.

Our story is not unique. The scientific challenge and opportunity that Paul Chu's colleagues relished was typical of the experiences of more than a thousand other researchers around the world. To be sure, Paul Chu was first; his research team discovered superconductivity in the exotic mixture of yttrium, barium, and copper oxides weeks before anyone else. Their advance was pivotal because it captured the imagination of the scientific world and altered the research goals of thousands of workers in dozens of countries.

But virtually every other discovery associated with the new material—the separation and identification of the 1-2-3 superconducting compound, the measurement of its unique electrical and magnetic properties, the determination of its beautiful and novel crystal structure, and the synthesis of dozens of new high-temperature superconductors of different compositions—was made independently and almost simultaneously in other labs worldwide. Everyone involved—in China, France, Great Britain, India, and Japan, as well as in North America at Argonne National Lab, Bellcore, Bell Labs, Berkeley, IBM, and Stanford, to name just a few of the first research teams to tackle "high-T_c" compounds—grappled with the same roadblocks to the unknown. My story, then, is only one example of the intense efforts and extraordinary excitement shared by researchers at more than a hundred academic institutions, government laboratories, and industrial research centers in Asia, Europe, and the Americas.

Few scientific discoveries are as dramatic as that of Paul Chu's superconductor, which was treated as a major international news event. Most scientific insight, including our imperfect, still-growing understanding of the strange new compound nicknamed "1-2-3," is gained by a gradual, day-by-day wresting of tiny bits of information that in time may form a coherent picture of some small corner of nature. The process of scientific research is often slow, repetitive, and exacting, but the rewards can be great: the exhilaration of discovery more than offsets the tedium.

Science is a human endeavor and my story would not be complete without reference to the unique personalities and dramatic personal interactions that are integral to the scientific process. All of the dialogue presented in the book is based on actual discussions, although in several instances the exact date or time of day has been shifted slightly for the sake of narrative continuity. My versions of conversations are taken in part from tape-recorded transcriptions made during March 1987, but portions of many conversations are recreated according to my notes and recollections. The result is an honest, but an admittedly personal, view of the memorable events leading up to the epic all-night meeting of the American Physical Society on March 18, 1987, when the gleanings from the first round of superconductor research were presented to the world.

R.M.H.

Prologue

January 1986

Superconductivity. The word is magical, just like the phenomenon. Superconductivity is perpetual motion on an atomic scale, the conduction of electricity without the slightest power loss—perfect conductivity. In the first months of 1986 superconductivity was the Holy Grail for a handful of scientists around the world. They sought to comprehend its fascinating physics and tap its potential applications.

But there was one seemingly insurmountable obstacle. Of the hundreds of superconducting materials found prior to January 1986, none would work without refrigeration to a few degrees above absolute zero temperature. (Absolute zero or zero Kelvin, usually abbreviated 0 K, is equal to minus 273 degrees Centigrade [written −273°C] and is the lowest possible temperature. Atomic vibration effectively ceases at 0 K.) Superconductor researchers were constantly struggling to find new materials with easily attainable high critical temperatures, or "high T_c," at which superconductivity commences. But, in spite of seventy-five years of research, the high-temperature record had risen only 19 degrees since the original discovery of 4-K mercury superconductivity by Dutch physicist Heike Kamerlingh-Onnes in 1911.

From 4 K the record had crept upward every few years, a few degrees at a time. But for more than a decade the highest temperature, observed in exotic alloys of the rare metal niobium, remained at about 23 K. A modest industry had arisen for the production and application of powerful superconducting electromagnets fabricated from these alloys. In recent years, however, many scientists had abandoned research on superconductivity. Wishful thinking about commercially viable high-T_c superconductivity was one thing, but hard data from the real world told another, less optimistic story. Many researchers assumed that further substantial advances were unlikely, perhaps even physically impossible.

But Georg Bednorz and Alex Müller, physicists at IBM's Zurich Research Laboratory, didn't give up quite so easily. Rather than study the traditional types of superconductors such as metals and alloys and the like, they followed a hunch and concentrated on oxides—chemical compounds in which the metal atoms are bonded to oxygen. To many it seemed an odd choice. Most oxides are insulators, with no appreciable electrical conductivity, though a couple of odd metal-like oxides had been shown to be superconducting. In 1973, for instance, David Johnson at the University of California, San Diego, observed superconductivity in a lithium-titanium oxide, and two years later Art Sleight at DuPont saw the same effect in an oxide of barium, bismuth, and lead. But those materials had to be cooled to a frigid 13 K, well below the 23 K of recording-holding niobium alloy materials. Few scientists expected higher temperatures from ordinary oxides.

But the two Zurich physicists reasoned differently; they guessed that superconductivity might occur in a small group of odd nickel- or copper-oxygen compounds that displayed peculiar electronic behavior at room temperature. Working without laboratory assistants and with little encouragement from their peers or employers, they mixed elements together in different proportions, cooked the mixtures in high-temperature ovens, and chilled the synthetic samples that emerged, looking in vain for the elusive effect. They labored for two and a half years, synthesizing sample after sample.

Elder partner Karl Alex Müller, with simultaneous senior appointments as professor of physics at the University of Zurich and fellow at IBM's Zurich lab, could work on just about anything he wanted. Sixty years old, with almost a quarter century of IBM service to his credit, Müller had paid his dues. But his younger recruit, thirty-seven-year-old West German Johann Georg Bednorz, had less freedom. Hired by IBM in 1982, he was paid a good salary to do company work on company projects, not to go off on some half-baked treasure hunt. By agreeing to collaborate with Müller, Bednorz was forced into the uncomfortable situation of secretly synthesizing potential superconductors while simultaneously carrying out his authorized lab duties. As sample after sample failed, the Zurich team's frustration grew.

Every scientific discovery, no matter how revolutionary, builds on earlier findings. The IBM Zurich breakthrough was no exception. In late 1985 Bednorz read a series of papers on unusual copper oxides that had been synthesized by Claude Michel and his coworkers at the University of Caen. In those papers the French workers noted some unusual metal-like electrical conductivity, a property not usually associated with oxides. The Caen group never tested for superconductivity. Bednorz did, and the material worked.

On January 27, 1986, a new era of superconducting science and technology began. Bednorz and Müller, using a variant of the materials synthesized by Michel, smashed the long-standing 23-Kelvin temperature record with a compound of barium, lanthanum, copper, and oxygen that at 30 K displayed the dramatic electrical resistance drop that is a key indicator of superconductivity. The younger Bednorz, frustrated by years of secrecy, was ready to tell the world immediately, but conservative Swiss-born Müller was more cautious. Their result would seem unbelievable to many and, in a field where many claims of high T_c had proven premature, Müller wasn't about to telephone the newspapers. Their scientific reputations were on the line. So Bednorz and Müller cautiously, painstakingly, repeated the experiments until they were absolutely convinced.

Three months passed before Alex Müller dared to show the extraordinary results to anyone. Finally, in mid-April, Müller handed a manuscript with a conservative and unassuming title, "Possible high T_c superconductivity in the Ba-La-Cu-O system," to his close friend Wolfgang Bückel, who was on the editorial board of *Zeitschrift für Physik*. Though widely read, *Zeit. Phys.* (common shorthand for the monthly) is by no means the most prestigious physics publication. The journal's advantage to Bednorz and Müller was the confidentiality ensured by their inside contacts on the editorial staff. Bückel, himself a well-known figure in the world of superconductor research, reviewed the paper. And Bednorz's boss, Eric Courtens, also a *Zeit. Phys.* editor, expedited its publication. Even with the paper accepted for publication, Bednorz and Müller did not present the results at any scientific meetings. In a decision that must still give nightmares to IBM executives, they did not even share their history-making data with company colleagues at the large IBM laboratories in New York or California where the findings and their extraordinary ramifications could have been studied in confidence. It took almost half a year more before the historic essay appeared, without fanfare, in the September 1986 issue.

Though no one realized it at the time, this publication marked a historic advance. Not only had Bednorz and Müller eclipsed the previous critical temperature record, but they had done it with a class of superconducting compounds that was all but unknown to workers in the field. By the late fall of 1986 the superconductor rush was about to begin.

Part I.

HOUSTON

Chapter 1.

Confirmation

November 1986

There are certain days in everyone's life that stand out against the continuum of routine events. Paul Chu remembers Thursday, November 6, 1986, as such a day.

Only a hint of dawn colored the darkened sky as Chu pushed through the heavy glass doors of the University of Houston's Science and Research Building and bounded up the stairs to his modest fourth-floor office. The stairway and corridors were deserted, and the early hour would give him uninterrupted time to catch up on some of the correspondence and other paperwork that had inexorably accumulated during the month he'd spent at the National Science Foundation headquarters in Washington, D.C. Chu's return to Houston was an escape of sorts, if only for a couple of weeks, from the pressures of science policy and funding decisions. It was a chance to see his wife May and their two young children, who had stayed behind in Texas. It was an opportunity to catch up on lab business, neglected for more than a month. And there might even be time to enjoy a tranquil hour or two preparing the garden for winter. Paul Chu needed those quiet times to keep his hectic life in perspective.

Reaching his office, Chu switched on the light and glanced at his functional metal desk to assess the damage. The stacks of mail that buried the desktop seemed painfully high after five weeks of neglect. He'd have to work fast to make even a small dent by the time his colleagues started to arrive. Without so much as a cup of coffee to spur him on, he settled himself in his swivel chair and began to work. As it turned out, he didn't get beyond the first few items that morning because there, near the top of the stack where his graduate student Li Gao had placed it the previous day, was a freshly copied, five-page article from *Zeitschrift für Physik*.

Chu could barely contain his excitement as he reread the title:

> "Possible High T_c Superconductivity in the
> Ba-La-Cu-O System."

Georg Bednorz and Karl Müller had set a world record. The article described the evidence for superconductivity at 30 K and the steps necessary to perform the experiment. Scientists in dozens of other laboratories had seen the same article; most had dismissed the paper as just another wild, unsubstantiated claim. But the Houston physicist thought differently. He was sure that this was the breakthrough he had been waiting for.

Paul Chu had labored for years on superconductivity and had searched in vain for the elusive high-temperature superconductor that would transform the esoteric phenomenon into a reality. It had been hard work, but it had also been fun. As his mentor Bernd Matthias had shown him, the study of superconductivity was as much an art as a science. The research required skill and dedication, but it also required a bit of luck.

Paul Chu's tortuous path to the Houston physics lab had prepared him to seize the lucky breaks when they occurred as well as to accept the inevitabilities of failure when necessary. Born in 1941 in China's beautiful southern uplands, he was given the name Ching-Wu. Though his family lived in the troubled province of Hunan, they survived the destructive Japanese invasion of the early 1940s only to find themselves embroiled in the three-year Chinese civil war that began in 1945. In 1948 Chu's parents, who were members of the Nationalist Party, were compelled to move south to the coastal city of Guangzhou as the Communists gained control of most of the mainland. Education for Ching-Wu and his five siblings was all but impossible during that turbulent time. By the summer of 1949 it was evident that the Nationalist cause was lost and in September the Chu family fled on a military transport to Taiwan.

On Taiwan Ching-Wu Chu began the science education that eventually led to an undergraduate degree in physics from Chengkung University in 1962. Like most other promising Chinese science students of the time, Chu hoped to pursue graduate studies in the United States. Ching-Wu's graduate science education began in 1962 at Fordham University in the Bronx, where he enrolled as a master's degree candidate, following in the footsteps of a Taiwanese friend and schoolmate. Having arrived in the United States, Chu began to use his Christian name, Paul. Fordham professor Joseph Budnick, who taught physics and studied the properties of magnetic materials and superconductors, encouraged Chu to study magnetic cobalt alloys for his master's thesis. Paul Chu excelled, building his own sophisticated experimental hardware at the modestly equipped laboratory.

Chu's educational odyssey continued with doctoral studies in San Diego at the beautiful University of California campus. It was there that he came under the spell of Bernd Matthias, the noted senior statesman of superconductivity. While a scientist at Bell Labs in the late 1940s and 1950s, Matthias had played a major role in understanding and improving superconductors. In 1961 he accepted a physics professorship at San Diego, while maintaining formal ties to Bell Laboratories as a consultant.

Of the many lessons Matthias taught Paul Chu about the scientific process, the most important ones were not to be found in any textbook. "Follow your hunches; use your intuition," he would say. He taught Chu to scan the scientific literature regularly and systematically for promising new materials, techniques, or theories that might help in understanding the mystery of superconductivity. He told Chu to listen to his dreams.

Matthias was instrumental in landing Chu his first job—a two-year postdoctoral fellowship at AT&T Bell Laboratories in New Jersey, working on superconductors and other useful compounds at high pressure. Those two years were intense and stimulating, but Chu sensed that in the huge New Jersey facility he lacked the freedom to set his own course. So he entered the world of academia, first as an assistant professor at Cleveland State University and then, in 1979, as a full professor at the University of Houston. With a modest budget and large ambition he gradually built a research team composed of graduate students and postdoctoral fellows. The enthusiastic young professor was especially adept at attracting gifted Chinese and Taiwanese researchers who saw in Chu a dynamic role model—as well as a teacher who could speak both English and Chinese.

Inspired by Chu's infectious good nature and his calm, Zen-like acceptance of the many failures inherent in the superconductor game, the lab team put up with his unconventional intuitive approach. Once, in 1982, Chu experienced a vivid dream in which sodium sulfide, an ordinary stockroom chemical, became a superconductor at high temperature. He galvanized the lab into action in a time-consuming search for the fantasized effect. The experiment was unsuccessful but Chu's enthusiastic group, which at the time included a bright young Taiwanese graduate student named Maw-Kuen Wu, never complained about his idiosyncratic style.

The outcome of a superconductor experiment is impossible to predict. As if to underscore this uncertainty, Chu's lab members would organize a betting pool to see who could guess the critical superconducting temperature of each new unknown compound. And guesses they were, in spite of decades of collective scientific training. Chu, ever the optimist, would almost always guess a T_c too high.

And, if the fun and adventure weren't reason enough, they knew in the back of their minds that Chu's visions might some day make them famous.

By the summer of 1986, after having devoted most of his scientific career to the search for superconductors, Paul Chu was becoming less optimistic. Time—and research money—was running out. He had privately resolved, in those hot, humid months of 1986, to abandon superconductors altogether if he couldn't achieve 30 K by 1989, the year of his next grant renewal. Now—suddenly—all those uncertainties were over; Bednorz and Müller had done it. Chu was electrified; he would set his entire laboratory to work at once on the project. He read and reread the paper and then walked across campus to the university library where he nervously passed the time searching for as much related information as he could find until the hour for his research team to show up for work.

By 9:00 A.M. the members of the group had arrived and Paul immediately called a meeting. His new orders were succinct: Stop everything and work on the La-Ba-Cu oxide superconductor! In practical terms it was a decision that would change the focus of their research for months to come.

There are four basic steps in the discovery of a new superconductor. First, obtain a sample, usually through a synthesis process reminiscent of creative cooking: mixing chemicals in careful ratios, grinding them up, and baking until done. In addition to the exact chemical mix, the researcher must determine the optimum temperature of synthesis, the correct gas atmosphere during synthesis, the annealing time at high temperature, and the cooling rate. Sometimes the mix-and-grind technique doesn't work and other complex synthesis procedures involving high-temperature liquids or gases are required. Experienced researchers know that while a single synthesis experiment may take only a few hours, the successful creation of a superconducting sample often requires weeks of patience and creativity.

The second step in establishing the existence of a superconductor is to determine the T_c, the "critical temperature" at which electrical resistance drops to zero. To accomplish this, thin electrical wires are attached to the sample, which is then lowered into an insulated bottle of liquefied gas refrigerant, usually liquid helium. Hoping to see the precipitous drop in resistance, the scientist charts electrical conductivity versus temperature. But a resistance drop by itself can be misleading. A faulty connection or a short circuit can trick the physicist into believing an experiment has been successful. Confidence in the resistance measurements is gained by repeating the experiment while the superconducting sample is subjected to a magnetic field. The T_c *always* drops to a lower temperature in a magnetic field, and scientists must verify that effect in their quest to confirm superconductivity.

The third superconductor test is the demonstration of the "Meissner effect," a subtle phenomenon related to a unique magnetic property of superconductors—the ability to "exclude" a magnetic field. Magnetic fields are all around us, generated by the earth, by permanent magnets, and by electrical devices. These invisible fields pass right through ordinary objects; that's why magnetic fields generated deep within the earth can guide ships and planes on the surface. But superconductors aren't ordinary objects. Magnetic field lines are pushed aside by superconductors. One startling result is that a permanent magnet will float, as if by magic, above a chunk of superconducting material that has been chilled to its critical temperature.

Physicists confirm the Meissner effect by measuring "magnetic susceptibility"—a property related to the degree to which magnetic field lines are distorted by the material. Even a tiny amount of a superconductor mixed in with other material will cause ripples in the magnetic field, and these ripples can be detected in the laboratory. Imagine a granular sample composed of 95 percent non-superconducting solids and only 5 percent superconductor. As temperature is gradually lowered below the critical temperature, the superconducting part of the sample will cause small but distinctive changes in the magnetic field. The sample probably will not show zero resistance because 95 percent of the material is not superconducting. But the Meissner effect reveals the presence of a superconductor, even if it only accounts for a few percent of a sample. For that reason scientists accept the presence of the Meissner effect as the most compelling evidence for superconductivity.

The fourth step in verifying that a material is a superconductor is isolation and characterization of the specific superconducting compound. Many superconductor samples consist of a mixture of more than one type of compound, or "phase." Each phase is characterized by its own distinctive chemical composition, atomic structure, and physical properties. Common table salt is a single phase containing sodium and chlorine. A glass of ice water has two phases of H_2O, one solid and one liquid. Most rocks are a cemented mass of several different phases (called minerals). A superconducting sample may also have several different phases cemented together, but usually only one of these phases is the superconductor. The physicist's job is to isolate and purify the correct superconducting material and then determine its chemical composition and atomic structure. With that information in hand, engineers can search for applications for the pure superconducting material.

Bednorz and Müller's paper reported the achievement of the first two necessary steps—they had synthesized samples and they had observed electrical resistance begin to decrease at the record high temperature of 30 K, with zero resistance at 12 K. But at the time of their discovery they did not have access to the sophisticated equipment necessary to demonstrate the Meissner magnetic effect, nor had they yet isolated the superconducting phase. Their samples, they had shown, were a fine-grained mixture of three different crystalline compounds; they assumed that only one of the three was the substance of interest.

Paul Chu and his coworkers agreed that their first step should be to reproduce the IBM Zurich discovery. Confirmation was not just a warm-up exercise; reproducibility was an essential (and publishable) part of superconductor science—some would even say the fifth essential step in proving superconductivity. For their synthesis experiments the Houston scientists needed the appropriate chemicals—compounds of lanthanum, barium, and copper—and a method to cook them in the correct proportions. Obtaining the chemicals was no problem. Bottles of powdered lanthanum oxide (La_2O_3), copper oxide (CuO), and barium carbonate ($BaCO_3$) were ordered immediately from chemical suppliers and would be delivered overnight. But the synthesis procedure posed a significant difficulty.

The Zurich article recommended a tricky synthesis technique known as "coprecipitation," in which the three metal elements—lanthanum, barium, and copper—are chemically dissolved in a solution and then precipitated to form an intimate, gelatin-like mixture of the elements. This amorphous, random mixture of atoms, Bednorz and Müller had found, could be baked in a high-temperature oven to create pellets of the new superconductor. Chu's group specialized in solid-state synthesis, a much simpler procedure in which dry powdered chemicals are mixed, ground, and baked. They had never tried the coprecipitation technique, and it was sure to take valuable time to learn this entirely new procedure. Under most circumstances, with a material of less importance, the logical approach would have been to try both techniques, carefully comparing the results and choosing the best method after several weeks of experiments. But this was no ordinary material. Time was of the essence if Chu's team hoped to beat other labs in the discovery of new and better materials. Which synthesis method should they use?

Research Associate Ru-Ling Meng, the synthesis expert who helped to run the lab during Chu's frequent travels, was strongly in favor of following the Zurich instructions to the letter. After all, she reasoned, it would be impossible to duplicate the IBM results without similar samples. Chu, on the other hand, leaned toward the simpler solid-state method. Meng, who had known Chu much longer than the other members of the Houston lab, did not hesitate to argue the point. A lively debate (in Chinese) ensued, intensifying as Meng emphasized the telling paragraph near the end of the IBM paper: "The way the samples have been prepared seems to be of crucial importance." Bednorz and Müller had already attempted solid-state synthesis at 1000° C and failed to produce any superconducting material. As Chu listened to Meng's arguments on that November morning he could not help contrasting her assured, confident style with that of the deferential, forty-year-old scientist he had first met in China eight years before.

In 1986 almost all Chinese scientists were either older than forty-five or younger than twenty-five; almost none were in their thirties. The Cultural Revolution had created a scientific void that stopped a generation of potential scientists from pursuing their goals. From 1966 to 1976 scientific research in China virtually ceased, while physics professors became farmers and farmers became university professors. Families were divided and careers were destroyed. For more than ten years scientific education was abandoned; a generation of Chinese youth was denied access to the knowledge on which a technological society depends.

Now there are two distinct age groups of "junior-level" research scientists in China. Thousands of scientists like Ru-Ling Meng, who were graduate students just before the Cultural Revolution began, are now in their forties, having wasted many productive years. Those who began their long scientific education after the Cultural Revolution are only now beginning to graduate with doctorates; they are in their early twenties.

These consequences of Mao's policies were very much in evidence in 1979 when Paul Chu spent several weeks at the Beijing Institute of Physics. He had been invited to help in the design of a superconductor research laboratory. In their intense efforts to catch up with the world and modernize scientific research in China, the Institute of Physics not only relied on visits from foreign experts, but they also sponsored many of the "junior" Chinese scientists to study in European and American laboratories. Thus Ru-Ling Meng, who was working at the institute at the time of Chu's China visit, returned to the United States with him and remained for two years as his research associate. In 1981 she rejoined the Beijing research center to transfer her new expertise and continue her research, but four years later she was given the chance to return to Houston, where she became a permanent resident and Paul Chu's full-time laboratory research associate.

There was no clear-cut answer to the synthesis debate, but Chu had to make a decision—coprecipitation or solid-state methods? In the end, as he so often did, Chu let intuition be his guide. He decided to try the easier solid-state approach first. At worst, if they failed, they would only waste a few days in the attempt. And if it worked, they would be able to produce samples much more easily than laboratories that relied on the coprecipitation procedure.

On Monday, shortly after receiving the new chemicals, Ru-Ling Meng went to the superconductor laboratory to begin the synthesis. The lab was in its usual chaotic state. A seemingly random array of scientific doodads littered the unimpressive twenty-by-thirty-foot room with its tan linoleum tile floor and cinder block walls. But Chu's superconductor lab was designed for efficiency, not looks.

Along one wall, to the right of the main doorway, is a row of white-topped desks and cabinets forming a thirty-foot-long lab bench, strewn with wires, tools, microscopes, and other materials used in the preparation of samples. Electrical resistance and high-pressure experiments are assembled there. Adjacent to the lab bench are two smaller tables, one with grinding supplies and the other with a high-temperature furnace for baking samples. Windows on the far side of the room overlook the Houston campus, but the view is largely obscured by banks of library-type metal shelves holding electronic gear, power cables, high-pressure fittings, and a myriad of long-neglected hardware that might someday find a use. To the left of the shelves, on the wall opposite the long lab bench, are doors that lead to the offices of Chu and his colleagues. The heart of the laboratory is a twelve-foot-wide framework of metal braces and copper tubing that stretches from floor to ceiling in the center of the room. That assembly supports the critical distribution of liquid helium refrigerant to experimental stations where superconductors reveal their cryogenic secrets. Ten or more physicists, each performing a different task, can work simultaneously in Chu's efficient, self-contained superconductor laboratory.

Meng began the synthesis by weighing precise ratios of the lanthanum, barium, and copper chemicals and mixing them in an agate mortar. She ground the pure white oxides together until the mixture had the appearance of the finest talcum powder. Meng then poured a layer of this powdered chemical stew into a steel piston-cylinder arrangement. Two steel pistons, about the diameter of a pencil and driven by the force of a hydraulic ram, compressed the powder into a dense oxide pellet. The operation was repeated five times. She placed all five pellets into a ceramic crucible and fired it to a bright orange incandescence at 950° C. The results of her efforts were five identical disk-shaped samples, each no larger than an aspirin. The pure white powders, under the transforming influence of the oven, had metamorphosed to unimpressive dull black masses, with granular surfaces and brittle-fractured edges.

There was no guarantee that all portions of all the disks had baked in exactly the same way, so the Houston group cut each disk into three elongated 1-by-1-by-3-millimeter rectangular prisms. Each of the fifteen resultant pieces was prepared for the all-important measurement of electrical resistance at low temperature. The researchers attached four slender platinum wires at equal intervals along the axis of each tiny piece, and each piece, in turn, was chilled to the incredibly cold conditions necessary to unleash the reported free flow of electrons.

The eager physicists, hoping to see signs of the abnormal behavior described by Bednorz and Müller, monitored electrical resistance as they gradually lowered temperature. At first the temperature-versus-resistance curve on their graph was routine. From temperatures of 80 K to 60 K to 40 K the resistance underwent a typical slight increase. But at 30 K they began to see the dramatic drop. Chu's intuition turned out to be right—the quick and easy solid-state synthesis had worked. And they had shown that superconductivity at 30 K was real!

Or was it? The temperature of superconductivity is affected by a magnetic field—the higher the magnetic field, the lower the critical temperature. Chu and his coworkers had to run the resistance measurements again, this time in the presence of a strong magnetic field. Sure enough, the critical temperature dropped. Good fortune had smiled on the elated Houston team, for only one other slice of the fifteen original pieces duplicated the first successful result. So tenuous was the effect that all fifteen runs might easily have failed. What would have happened if none of the samples had worked that day? How long would the team have persevered before they abandoned the project for other, slower sample preparation methods or other compositions? Chu himself wasn't sure. But once he saw the superconductivity, and believed it was real, there was no turning back. And with the sure knowledge that solid-state synthesis worked, Paul Chu thought he might just have the edge he needed to discover even better superconductors.

The following days were filled with a succession of repeat experiments, with similarly spotty results: out of each synthesis run only a few black chips became superconducting at low temperature. The great majority of samples showed nothing at all. Some samples would give clear signs of superconductivity, only to lose the effect after three or four days. All of the Texas-based researchers were absolutely convinced that the IBM results were valid, but they had yet to stabilize the phenomenon. They had duplicated the IBM experiments, but it was apparent that only a small fraction of each synthetic sample was composed of the superconducting stuff.

What was the problem? When several oxides are baked together they will not, in general, form just one type of crystal. Most common rocks, for instance, contain half a dozen different oxide minerals. The IBM recipe, with a mixture of three oxides, was like a rock with three different coexisting minerals. Perhaps only one of the three was superconducting, or perhaps superconductivity was a consequence of the special atomic arrangements that occurred at the boundaries—the "interfaces"—between two different compounds. Chu wasn't sure. But he and his coworkers were sure that the superconducting signal could be improved, if only the chemical compositions and synthesis conditions for those three coexisting phases could be determined. By the third week in November the Houston team knew they would have to change tactics and search for a new, improved chemical mix. But this next series of experiments would have to proceed without Chu.

Paul Chu headed north for scientific conferences and National Science Foundation business, but his absence wouldn't slow down the effort. Everyone in the Texas laboratory was used to Chu's running things by telephone. His group had learned to be ready for his calls day or night. They might be asked to give a report on the day's progress or to take instructions for another experiment. It was a smooth-running operation as long as the telephone was working.

First on Chu's mid-November agenda was the 31st Annual Conference on Magnetism and Magnetic Materials (the "3M" meeting), held from Monday, November 17, to Thursday the twentieth at the Hyatt Regency Hotel in Baltimore. The yearly affair attracted an eclectic mixture of physicists, materials scientists, and engineers, all of whom shared an interest in magnets, and a modest number who specialized in superconductivity. Paul Chu was scheduled to give a Tuesday afternoon talk on the physical properties of iron oxide (not a superconductor), but superconductors were foremost on his mind. The general mood of other superconductor researchers was subdued, and a number of scientists were openly pessimistic about the future of their chosen field.

Chu's first impulse was to tell his friends the good news about the exciting new lanthanum-barium-copper oxide. He wasn't about to make a formal announcement just yet, but he could tell people individually. He spotted Harold Weinstock, an acquaintance from the Air Force Office of Scientific Research, and began an animated description of the latest spotty results. He reviewed the discovery of Bednorz and Müller and described his group's tentative confirmation. There was much left to be settled, Chu acknowledged, but the signals were there. Superconductivity at 30 Kelvin exists, he insisted.

"I'm not sure I believe it," was Weinstock's wary response. "Be very careful before you make any public claims." Chu was disappointed at the skeptical response, but he had to admit that the cautious viewpoint was not unreasonable. Most physicists had abandoned the notion of high-temperature superconductivity (in this context "high-temperature" means anything above 23 K). Why should anyone believe the inconclusive and sometimes nonreproducible Zurich or Houston data? He reluctantly resolved to keep his information confidential until better results were available.

From Baltimore Paul Chu headed to Washington and the headquarters of the National Science Foundation where he was serving a one-year term as NSF program director in the Division of Materials Research. He was responsible for allocating limited research funds to the large pool of deserving scientists. And the job wasn't getting any easier. More and more gifted researchers were struggling for a piece of the government pie, while glamorous big-bucks items like the newly proposed six-billion-dollar superconducting supercollider—a facility where a relative handful of high-energy physicists could search for subatomic particles—put severe demands on the government's science budget. With progress in new superconductor science and technology discouragingly slow, it was ever more difficult to rationalize the annual budget increases required just to keep the handful of active laboratories going; it was nearly impossible to fund any of the worthwhile new proposals that arrived monthly. Ironically, as Chu struggled with the funding shortfall in Washington, the Houston lab was rapidly converging on a solution to the money dilemma. Within a few months hundreds of scientists would be demanding funding for high-T_c superconductor research, and President Reagan and the United States government would be more than happy to oblige for this hot new topic.

This rosy future was hidden from view in November 1986.

It seemed that individual materials researchers, and along with them individual initiative in research, were losing out to big-project science. That crisis would be a key subject of conversation at the Thursday, November 20, meeting of the NSF advisory panel on scientific computing. The panel would have to decide how limited funding for computing should be allocated to deal with the growing demand. For most scientists the computer is an essential piece of experimental hardware. Computers enhance almost every phase of scientific research; they control the day-to-day operation of laboratory equipment, they store and manipulate data, and they aid in the preparation of manuscripts, figures, and tables for publication. And for some scientists, like theoretical physicist Arthur Freeman of Northwestern University, computers *are* their research laboratory.

Art Freeman uses supercomputers to understand matter. Atoms and their interactions can be modeled with complex mathematical equations. Calculations that mimic atoms in nature are much too complex to complete by hand, but giant computers can do the trick. The first tests of these computer methods merely duplicate measurements of the laboratory experimentalist; computers might confirm that a certain atomic structure is the most stable, or that certain physical properties are observed. Theory helps to explain observation. Eventually Freeman and his colleagues may be able to predict the properties of potentially useful materials—substances that have never been made in the laboratory. If that happens theorists could point experimentalists to new, economically important compounds.

Art Freeman gladly served on the NSF panel because his research future largely depended on access to supercomputers. But he also enjoyed the chance to talk with Paul Chu regularly about their mutual interest in superconductors. Chu almost told Freeman about his recent confirmation of the IBM result. But instead, when asked about his latest lab work, he just gave Freeman an enigmatic reply: "Something's going on. I'll tell you later when I have more." Art Freeman would know soon enough.

While Chu sat in a Washington office building and pondered the state of science in America, he left his Houston lab in good hands. In addition to Ru-Ling Meng, postdoctoral fellow Pei-Herng Hor, graduate students Jeff Bechtold, Li Gao, Zhi-Jun (Peter) Huang, and Janis Vassiliouv, undergraduates Daniel Campbell, Theresa Lambert, and Andy Testa, and visiting research associate Ya-Qin Wang from the People's Republic of China were all hard at work on the project. Experiments often lasted well into the night, with hasty midnight dinners at the local Burger King.

The Houston researchers gradually zeroed in on the best synthesis procedures. They soon discovered that baking the chemicals in a vacuum produced a useless electrical insulator —the opposite effect from the one they were looking for. Evidently oxygen had to be present in the atmosphere for the creation of superconductivity. Synthesis at temperatures higher than 950° C also failed, so the group thereafter stuck to the 900°–950° C range. Knowing these parameters of synthesis, the Houston team next varied the composition of the superconductor itself by mixing different proportions of lanthanum, barium, and copper chemicals. The original Zurich sample, they knew, was a mixture of three unidentified phases. In such a mixed-phase, granular sample there could be no guarantee that a continuous electrical contact of conducting grains spanned the entire specimen. By changing the proportions of the starting chemicals they could change the relative amounts of the three resulting compounds. If they could identify and isolate the pure superconducting phase, a much better superconducting signal should result.

The original Zurich superconductor recipe had a lanthanum to barium to copper ratio of about 4 to 1 to 5. Chu's group, after their initial ten days of trials, began to vary this ratio and soon found improved results with a smaller proportion of copper. Specimens of that new composition were prepared on November 25, two days before Thanksgiving. Chu, who had by that time returned home to participate in the testing, joined the other physicists who crowded around the graph recorder. Everyone hoped to see the electrical resistance drop, but no one was prepared for what they saw: a sharp and clear superconducting transition at 73 K—the highest temperature ever. The new sample was a breakthrough. Or so they thought.

Paul Chu instructed his team to go home and get a good night's rest. They would all gather to repeat the experiment the next day. Success, Chu thought to himself, would be the best Thanksgiving gift he could imagine.

But what Chu's group hoped would be a breakthrough turned out to be just another failure. On Wednesday, November 26, all signs of superconductivity were gone. What was going on? The black chip had remained in air overnight and had somehow changed. The sample must have reacted with something in the air and its properties were altered. Paul Chu believed that the 73-K signal was real, but there was no way that he could publicize his result. No one would believe their inconclusive evidence. The history of superconductivity research is littered with unsubstantiated claims, false leads, and disappointment. And Chu could not forget his own involvement in the copper chloride controversy.

In the mid-1970s Chu had worked with an eccentric visiting Russian physicist, Alexander Rusakov, on the low-temperature, high-pressure properties of copper chloride—CuCl. Working at Cleveland State, the two researchers saw repeated spectacular anomalies in magnetic properties at temperatures as high as 200 K—anomalies that suggested that copper chloride might be superconducting. When the time came for Rusakov to return home, Chu urged his Soviet colleague to use restraint—more work was needed. But Rusakov soon announced that CuCl was a high-temperature superconductor, a result that made international headlines in 1978.

The CuCl controversy raged, as scientists at Bell Laboratories convincingly demonstrated that *pure* CuCl was not superconducting. Paul Chu countered with an alternative theory: perhaps superconductivity only occurred in impure CuCl as a result of unusual electronic behavior at the boundaries between CuCl and impurities. Chu's concept of "interface superconductivity" has never been proved, though it is still a favorite theory of his.

Whatever its origin, the previous day's observation of superconductivity at 73 K was only a teaser. There was still much to be done. Chu recognized that he would have to isolate and identify the different phases in his samples if significant progress was to be made. "X-ray powder diffraction" was the answer.

When a crystalline powder is bombarded with a finely focused beam of x-rays, the atomic layers of the crystal diffract x-rays at preferred angles. The set of angles at which a compound diffracts x-rays is like a fingerprint; each crystalline compound has its own distinctive pattern. But none of Chu's researchers was skilled in x-ray crystallography. Ru-Ling Meng had attempted to analyze an x-ray pattern, but the complex sequence of peaks and valleys seemed indecipherable. Trying to sort out two or more unknown phases in one powder pattern is something like trying to resolve two fingerprints smeared on top of each other. Chu's group needed help.

Fortunately, the laboratory adjacent to the fourth-floor superconductor research area is the domain of Professor Simon Moss and his crystallography team. One of Moss's doctoral candidates, an eager and talented student named Ken Forster, had expressed interest in superconductor research. In late November Paul Chu invited him to x-ray and analyze the La-Ba-Cu oxide samples. Forster, knowing his advisor's proprietary nature, was properly careful to ask Moss's permission to use the x-ray laboratory's state-of-the-art powder diffractometer. Moss thought the new superconductor was just another false lead, but he gave his approval—as long as it didn't affect the pace of Forster's thesis research. Within a few days over the Thanksgiving vacation, Forster had obtained well-resolved x-ray powder patterns.

Following their January 1986 discovery Bednorz and Müller had almost immediately realized that they had a multiple-phase sample. They, too, had attempted to identify the different crystal structures that made up the mixture with powder x-ray diffraction. They eventually identified three coexisting crystal phases: a copper oxide, an unidentified barium-rich compound, and a third material with a distinctive "layer-type perovskite-like phase, related to the K_2NiF_4 structure." The ratio of elements in this latter structure quickly earned it the nickname "2-1-4." In their *Zeitschrift für Physik* article, Bednorz and Müller noted that this new layered-structure phase with the 2-1-4 potassium-nickel-fluoride atomic arrangement was the "apparently superconducting" phase.

Ken Forster's powder patterns showed the same thing. They demonstrated a clear connection between the amount of the 2-1-4 phase and the superconducting response. Moss and Forster concluded that Chu's group should focus on that composition.

Thanksgiving Day is an international celebration at the Houston superconductor laboratory. Every year Paul and May Chu invite all of his lab personnel along with a number of Chinese students and other friends to his spacious home in the elegant Lakes of Fondren section of southwest Houston. The midday feast, prepared by the Chus for thirty or more guests, was a traditional American turkey dinner with all the trimmings. As if that wasn't enough to do, the Chus followed the American celebration with a multicourse Chinese meal in the evening. Turkey and cranberries were supplanted by crunchy jellyfish, spicy crab legs, and other Oriental delicacies. With so much to get ready, Thanksgiving was one of the few days that Paul Chu was in Houston but not at the laboratory, though the lab was no doubt on his mind.

Even if there was a slight undercurrent of disappointment at the disappearance of the 73-K signal, there was also a renewed sense of optimism at the promise of achieving high-temperature superconductivity. It had been only three weeks since they had so abruptly changed their research directions, and they had already learned a great deal. They had confirmed the work of Bednorz and Müller: a new La-Ba-Cu oxide superconductor had been found. They had discovered that the sample's oxygen content, which could change merely by exposure to air, had to be carefully controlled. They had demonstrated that the ratio of lanthanum to barium had a strong effect on the stability of T_c, the superconducting transition temperature. And they suspected that a composition enriched in lanthanum relative to barium and copper was best.

As they ate and talked, the physicists knew that discovery of superconductivity at 73 K—a discovery undreamed of a few weeks before—was within their grasp. Paul Chu sensed that success was only a matter of time. And he decided it was time to try something new.

Chapter 2.

Pressure

December 1986

Many people ignored the first formal journal announcement of Bednorz and Müller's work. Some scientists assumed that any valid IBM breakthrough accepted by a journal in April would have become common knowledge by the end of the year and that, conversely, the lack of any such rumors meant the results must be wrong: another premature announcement. Many scientists were too busy with ongoing research to follow the obscure lead. But soon, thanks in large measure to workers from Tokyo and Houston, the IBM discovery would be the talk of the superconductor world.

Paul Chu was not prepared to compete with IBM, Bell Labs, or other high-powered research corporations that were ideally suited to develop commercially the new superconducting compound. As soon as those laboratories learned of the Zurich discovery, he reasoned, they would concentrate on the obvious and important things: synthesis of perfect single crystals; measurements of the subtle Meissner magnetic effect; measurements of electrical current-carrying capacity; and measurements of the effects of strong magnetic fields (which cause T_c to drop). Those and other experiments were critically important to the scientists who characterized the samples and the engineers who would incorporate the new superconducting materials into their company's products.

But Paul Chu is not a conventional thinker and he wasn't interested in doing the conventional things. Chu's claim to fame in the world of superconductors was high-pressure research. Time and again over the previous two decades he had subjected superconducting samples to immense pressure of thousands of atmospheres in order to measure the shift of T_c with pressure. In some cases T_c went up slightly; in others it went down. Chu's objective in these experiments was not to find a superconductor that worked well at high pressure; no one needed such a material for any practical applications. But the high-pressure experiments always provided some insight on the reasons for superconductivity. Chu hoped that his high-pressure experiments would point the way to new compositions with better superconducting properties at room pressure. Even so, with years of experience under his belt, Chu was totally unprepared for the surprise of lanthanum-barium-copper oxide.

Measuring superconductivity at high pressure is a technical feat. The difficulty of measuring electrical resistance at a temperature near absolute zero is compounded when the sample, the electrical wires, and the thermometer have to be encased in a high-pressure device. The basic principle of high-pressure research is simple. Pressure is defined as a force per area; large pressures result from large forces acting on small areas. In Paul Chu's pressure device he placed the entire sample assembly in a slender Teflon tube, ¾ inch long and ¼ inch in diameter, which he filled with liquid and sealed. This Teflon "cell" assembly fits snugly in a 1-inch-diameter metal cylinder which guides two carbide piston rams. Tons of force can be applied on the narrow ends of the sample-holding Teflon cell, which then deforms and compresses. The deforming Teflon in turn transmits a uniform pressure through the liquid medium to the sample.

On a cloudy, early December morning the researchers prepared a carefully selected sample of the La-Ba-Cu oxide compound for its high-pressure ordeal. That specimen had a reproducible room-pressure resistance drop that began at 32 K. With sample assembly completed, the experienced lab team sealed the Teflon cell and inserted it into its steel vise. Pressure was raised to ten thousand atmospheres; it was up to the hardware. The anxious physicists could only watch the telltale chart as the temperature dropped lower and lower. They could hardly believe the first data. Resistance began to

drop at close to 40 K. The critical temperature had skyrock-
eted; Chu had set his first T_c world record. Never before had
such a fantastic pressure effect been observed in a superconductor. But of course they were dealing with a new class of
superconductors and it was becoming clear that conventional
wisdom no longer applied.

Chu and his coworkers immediately recognized the implications of their high-pressure work. If pressure raised T_c dra-
matically, then it was likely that higher T_c could also be found
at room pressure. The trick would be to find a new chemical
composition with smaller atoms that mimicked the squeezing
effect of pressure. The resulting compound would have a T_c
even higher than the Bednorz and Müller superconductor—
under ordinary pressure. A new round of synthesis experiments was about to begin.

With four weeks of data in hand, Paul Chu was at last certain
of his results. He decided to tell the world. Though only recently returned from his Washington and Baltimore trip, Chu
flew north once more, this time to Boston, on the morning of
December 4. More than 2500 scientists had gathered in cold,
wintry Beantown for the annual fall meeting of the Materials
Research Society, held during the first five days of December.
Throughout the week—morning, noon, and night—as many
as a dozen concurrent sessions focused on every imaginable
type of material: cements, radioactive waste, polymers—and
superconductors.

The December 4 symposium on superconducting materials
included Paul Chu as the last of only four scheduled speakers.
He began his afternoon presentation by describing some recent work on a different type of oxide superconductor, but
toward the end of his allotted twenty minutes he mentioned,
almost in passing, the resistivity data that supported the historic Bednorz and Müller claim. Chu was still in doubt regarding the chemical makeup of the superconducting phase,
and he did not reveal his group's preliminary findings of the
surprising high-pressure results, which had yet to be repeated
and interpreted. Even so, it was the first public confirmation
of the Bednorz and Müller discovery. The audience had much
to think about and the discussion period following the formal
presentation was a memorable one.

Amid the flurry of questions on Paul Chu's announcement there was one completely unexpected response. Dr. Koichi Kitazawa of the University of Tokyo's Department of Industrial Chemistry asked to make a statement. To be more accurate, Kitazawa upstaged Chu.

The Japanese scientist appeared to be excited as he quickly moved to the front of the hall. Kitazawa had earlier given his own invited talk on oxide superconductors without the slightest hint of Japanese work on the new La-Ba-Cu oxide compound. But *after* Chu broke the news Kitazawa announced that the University of Tokyo group had been working on the new compound since October. They had already performed an extensive series of experiments on the new superconductor. He told about magnetic and resistance measurements and illustrated his remarks with two carefully prepared viewgraphs. Kitazawa also hinted that the new phase had almost been identified and isolated.

It was extraordinary news. It meant that the Bednorz and Müller discovery of La-Ba-Cu oxide superconductivity, a discovery discounted or ignored by many physicists, had been duplicated by both Chu's group in Houston and Kitazawa's group in Japan. The confirmation electrified the audience. Many of those present had been unaware of the earlier IBM work, and others had dismissed the findings. But, with this new corroborative evidence, there could be little doubt that research on superconductivity was about to enter a new phase.

Even as Chu and Kitazawa were confirming the Zurich discovery, the first signs of interlaboratory tensions and jealousies—emotions that surfaced throughout the superconductor race—were being felt. Many people in the audience assumed that Kitazawa had purposefully avoided talking about the high-T_c work in the hopes of gaining a head start on rivals. It appeared to most attendees that the Houston remarks had forced the Japanese hand; Kitazawa's subsequent announcement, it seemed to them, was a hasty (some thought clumsy) attempt to gain credit after the fact.

In truth, Kitazawa had planned to present his results all along—but a day later. Upon arriving at the MRS meeting early in the week, the Japanese researcher told several people, including Stanford physicist Theodore Geballe, about the latest Tokyo work. In an unusual but logical move Geballe, who was MRS chairman for a Friday superconductor session, invited Kitazawa to give a special report. A lecture slot had become available in the Friday superconductor session due to the illness of Santa Barbara theoretical physicist Douglas Scalapino who, coincidentally, had just returned from a trip to Kitazawa's Tokyo laboratory. In mid-November the Japanese were intent on secrecy—so much so that when Scalapino lectured in Tokyo on oxide superconductors he was not given the slightest hint of the new research. But by the first week of December there were no such reservations.

During the night of December 4 Kitazawa phoned his laboratory for the latest results. In the course of the December 5 talk the Tokyo representative was thus not only able to show resistance and magnetic data that confirmed superconductivity in the Zurich material, but he also revealed for the first time the identity of the new high-T_c phase. With a combination of synthesis and x-ray experiments the Japanese had proven beyond a doubt that 2-1-4 was the superconductor.

Kitazawa believed that the new superconductor's atomic structure might provide a key to its superconductivity. The IBM superconductor is a copper compound with a layered atomic structure identical to that of potassium-nickel-fluoride. Kitazawa explained at the Boston meeting that an analogous oxide compound can be formed if the large metal atoms—lanthanum and barium—replace the potassium, copper substitutes for the similar-sized nickel, and oxygen proxies for fluorine. The resulting 2-1-4 compound, $(La,Ba)_2CuO_4$ with about five times more lanthanum than barium, is the superconductor. The most distinctive feature of the structure is the layering of copper and oxygen atoms. Two logical conclusions could be drawn from Kitazawa's presentation: in the search for better superconductors, first, synthesize materials with the 2-1-4 structure and, second, include copper as the smallest metal atom.

Kitazawa's revelations in Boston prompted several laboratories to shift into high gear. Excited scientists at Bell Laboratories invited Kitazawa to give a special lecture the following week at their high-powered research center in Murray Hill, New Jersey. By December 14 they, too, had duplicated the IBM findings. Theodore Geballe invited Kitazawa to lecture at Stanford, where the Japanese scientist presented Geballe with a copy of the just-completed phase identification manuscript. The Stanford group immediately set to work on preparation of thin films, a key step in computer applications of superconductors. Scientists at Argonne National Laboratory, Bell Communications Research, Berkeley, Northwestern, Stanford, and Westinghouse in the United States, as well as at the Institute of Physics in Beijing and several Japanese universities, also jumped on the superconductor bandwagon as soon as they heard the news from Tokyo. Dozens of researchers around the world had joined the game. High T_c in oxides, a concept that many had scorned a month before, had suddenly become a major opportunity for materials scientists.

With so many laboratories entering the superconductor game, Paul Chu knew he would have to have all the help he could muster to stay competitive. On December 4, in between scientific lecture sessions, he met with his friend and former Houston student Maw-Kuen Wu, who had become an assistant professor at the Huntsville campus of the University of Alabama two years before. Wu had always been willing to try new, risky things, so Chu asked for his help in synthesizing and testing potential new superconductors. In spite of Kitazawa's results, Chu had a hunch that the next big T_c breakthrough was not going to be a 2-1-4 compound. The transient superconducting signals that Paul Chu's group had seen above 70 K only occurred in "dirty" samples that weren't exactly on the 2-1-4 composition. Something else was causing those effects, but there was no way of guessing what it would be. They would just have to try dozens—maybe even hundreds—of new compositions. Wu enthusiastically agreed to join the hunt.

The theoretical study of superconductors, Chu knew, might also help in the search for new, improved superconducting compounds. No one was sure what made the IBM material work, but a theoretical physicist might be able to unlock the secret. Paul Chu recalled the recent November meeting of the NSF computer panel at which Northwestern University physicist Art Freeman had expressed keen interest in any new developments in oxide superconductors. Freeman's specialty was the computer calculation of electronic structure in crystals—a field that might shed light on the causes of superconduction in 2-1-4 materials. Once again the Houston physicist drew on his powers of persuasion, but Freeman took little convincing; he was delighted to join the ever-expanding Chu effort. In exchange for Chu's confidential updates on the properties, composition, and structure of any new superconductor, Freeman would perform calculations of the material's electronic structure. Those complex computations might reveal the nature of the superconducting mechanism and thus point the way to new compositions and structures.

Houston crystallographer Simon Moss, in spite of his former misgivings, also joined the converted. The Japanese revelations at the Materials Research Society meeting finally convinced him that Chu's work deserved support. Paul Chu welcomed the help, though he was annoyed at reports that Moss had been making offhand remarks about "sloppy science" in the Houston superconductor lab. But even though friction existed between the two University of Houston professors, Moss's student Ken Forster was encouraged by both to continue examining new samples as they were produced.

With more and more scientists around the world getting into the act, Chu also knew he should go to press. He returned to Houston during the second week of December and stayed just long enough to dash off a paper to *Physical Review Letters*, known as *PRL*, perhaps the most prestigious journal in the physical sciences. Chu's paper related his group's confirmation of the IBM findings, then emphasized the dramatic new high-pressure results. His title, "Superconductivity above 40 K in the La-Ba-Cu-O compound system," would make people sit up and take notice. Only the most important and newsworthy reports make it through the rapid, though rigorous, *PRL* review process, but Chu knew he had a winner. The manuscript was sent off on December 12.

Scientific lectures provided Chu with another forum to present his results. Gary Vezzoli, an old friend of Chu's from Fordham University graduate school days, had asked Chu to give a plenary address at an Army-sponsored conference on electro-optics held on December 15 and 16 at New Jersey's Picatinny Arsenal. The original lecture topic was to have been "Materials Processing in Space," a subject related to a NASA project that Chu directed in Houston. But the extraordinary events of early December led to a quick revision of the topic to "High-Temperature Superconductivity."

As Paul Chu labored to mobilize his extended laboratory and publicize his team's results, the Houston-based physicists were adding almost daily to their store of high-pressure data. Throughout the first two weeks of December measurements were repeated at several pressures for a range of La-Ba-Cu oxide compositions. The results were almost always the same. The largest pressure effect ever observed in a superconductor drove T_c above 40 K in sample after sample.

While the Houston group was busy setting high-T_c records, Maw-Kuen Wu and the Alabama researchers at Chu's bidding took another tack. The extraordinary high-pressure results had inspired the Houston physicist to bold thinking. High pressure, which caused the sharp increase in T_c, also forces atoms more closely together. But different atoms squeeze together, or "compress," different amounts. Chu knew that the barium atoms, the largest metal atoms in the 2-1-4 compound, were much more compressible than either lanthanum or copper. The high-pressure result might imply that decreasing the size of the barium relative to the other atoms caused T_c to rise; perhaps the same effect could be induced at room pressure by simply substituting a slightly smaller element for barium. One obvious choice was strontium, barium's next-smallest neighbor in the periodic table of the elements. So during the first week of December Chu assigned Wu the task of synthesizing the new 2-1-4 compound, $(La,Sr)_2CuO_4$. And, indeed, by December 14 Wu had created a 39-K room-pressure superconductor in the new oxide system.

Even without the striking high-pressure results, the strontium-for-barium substitution was attempted by several other labs. These odd new 2-1-4 copper compounds were unexplored territory and any simple chemical substitution was worth a synthesis experiment. Strontium and barium are elements that behave similarly in natural minerals, so why not in synthetic superconductors? In the last weeks of 1986 at least four other groups independently pushed T_c to new room-pressure records. Researchers at IBM, University of Tokyo, Bell Labs, and Beijing's Institute of Physics, as well as Houston, all did the strontium synthesis and all observed superconductivity at temperatures above the 30 K of the barium compound.

If strontium worked so well, would the next smaller element—calcium—work even better? The Houston and Huntsville researchers spent the last half of December synthesizing, squeezing, and measuring electrical properties of dozens of 2-1-4 samples, with varying proportions of lanthanum, barium, strontium, and calcium, as well as different synthesis conditions and cooling times. They found that lanthanum-strontium compounds (with these two metals in a 9 to 1 ratio) consistently gave the highest T_c, whereas the lanthanum-calcium mixtures had much lower critical temperatures.

As they gained skill in preparing the 2-1-4 superconductors, and as they obtained purer samples, pressure shifts even greater than the one reported to *PRL* were observed. During the third week of December Chu's lab team synthesized a sample of pure $La_{1.8}Ba_{0.2}CuO_4$ that displayed a reproducible high-pressure onset temperature of superconductivity near 50 K. It was another world record. Several of these samples also produced tantalizing transient effects at temperatures above 70 K, just as the pre-Thanksgiving specimen had done.

By the last week in December the popular news media were receiving the first hints of the superconductor mania that was soon to come. On December 27 the *People's Daily* of the People's Republic of China carried a front-page story on high-temperature superconductivity of La-Ba-Cu oxides. The story noted a possible 70-K effect that sounded to Chu just like the Houston results. Had the Beijing group, who claimed they had produced "the highest-T_c superconducting material," really seen the 70-K effect? Had they stabilized the phase? In any case, it seemed important that the latest Houston achievements be publicized.

Chu quickly prepared a second manuscript on the pressure effect, this one entitled "Superconductivity at 52.5 K in the lanthanum-barium-copper-oxygen system," and sent it to the weekly periodical *Science*. The report focused on the best Houston results for single-phase, 2-1-4-type material. A substantial portion of the paper concentrated on the variation of critical temperature, T_c, with pressure. As superconducting specimens are cooled the electrical resistance does not drop to zero all at once. Thus there are several different critical temperatures associated with any given experiment. The onset temperature, designated T_{co}, may be defined as the highest temperature at which the resistance starts to drop. The initial drop may be very slight and gradual, so T_{co} is sometimes hard to pinpoint. Easier to define are the midpoint temperature, T_{cm}, at which resistance has dropped to half its normal value, and the zero-resistance temperature, T_{cl}. The Houston group demonstrated that all three types of T_c increase with pressure.

Even with Chu's provisos, the 52.5-K value for T_{co} of the paper's title seemed a bit optimistic to some readers when the publication appeared a month later. Midpoint temperatures were closer to 35 K, and the zero-resistance temperature was never pushed above 30 K. But scientists agreed that the pressure shift was remarkable.

Chu's claim of the world's record high T_c was also the subject of a Houston press conference held on the morning of December 30. Since the *Science* paper had only just been submitted, Chu emphasized the earlier experiments, reported in *PRL,* with the 40-K onset T_c result. The 50-K and 70-K effects were only mentioned as intriguing possibilities. The Houston publicity effort was not wasted. Superconductivity made the front page of the *New York Times* on the last day of 1986, in an article by Walter Sullivan, the dean of the *Times*'s science staff.

Some physicists feel that Paul Chu was not given fair treatment by the New York newspaper on at least two counts. First, although Paul gave full credit to Bednorz and Müller and many other superconductor researchers in his press conference, the *Times* article made little mention of the IBM Zurich efforts. The headline and lead paragraph referred only to work at Houston and Bell Laboratories; Zurich was not mentioned until well into the article on page thirteen. To several scientists it appeared that Chu had tried to claim the discovery as his own.

Second, the importance of Chu's original contribution was distorted. In preparing his story shortly after Chu's announcement, Sullivan phoned Bell Labs for background information. By coincidence, unknown to Sullivan, on the previous day, December 29, Bell Laboratories had submitted a report of their discovery of "Bulk superconductivity at 36 K in

$La_{1.8}Sr_{0.2}CuO_4$" to *Physical Review Letters*. Not only had the Bell Labs group achieved high critical temperatures in the new La-Sr-Cu system, but they had also observed the Meissner effect, demonstrating beyond any reasonable doubt that the high T_c was real in 2-1-4 compounds. The New Jersey facility had not yet publicized these superconductor activities and they were evidently surprised by Sullivan's call. The *Times* reporter was asked to wait while Bell's press office and legal staff decided what to do.

Shortly thereafter the Bell Labs scientists returned Sullivan's call and told the reporter about their three-week-old efforts on the new compounds. Unfortunately, they were interpreted as saying that the Bell Labs superconductivity experiments "enjoyed an advantage [over Chu] because high pressure is not necessary," a comment that at best distorted the importance of the Houston results. As the Bell scientists knew, the Houston high-pressure experiments were never meant to substitute for room-pressure superconductors. Chu's high-pressure experiments pointed the way to new superconductors. It was a case of comparing apples and oranges.

But in retrospect the details of the *New York Times* article are much less significant than the fact of its publication. The world was being primed for the biggest discovery in materials research in decades.

The Christmas holiday at the Houston superconductor lab was usually a time for family and friends and a hiatus—or at least a slowdown—in laboratory experiments. But for the Houston and Huntsville researchers the last week of 1986 was a period of intense effort. Maw-Kuen Wu and one of his graduate students, Jim Ashburn, flew to Houston on December 27 with their best strontium-bearing samples in hand. High-pressure studies on those specimens were the next order of business for the Alabama researchers, while the Texas-based team continued their synthesis and testing of 2-1-4 oxides in the La-Ba-Cu system.

New Year's Eve provided the opportunity for a grand celebration. After an exhausting day of experiments Chu and Wu, along with an entourage of students, postdocs, and spouses, went en masse to ring in 1987 at Feng Ling, their favorite Chinese restaurant. It was a marathon affair, with a dozen different main dishes to share. They arrived just before midnight and the feast lasted well into the first hours of the new year. The researchers ate and drank. As they celebrated they talked of their strategy for the coming months. And when the dinner was over several of the scientists started off 1987 right by returning to the physics lab to try out new ideas born of the midnight revels.

Chapter 3.

The Yttrium
Breakthrough

January 1987

Within eight weeks of the time Paul Chu first learned of the Bednorz and Müller 30-K record he had twice set his own record high temperature—first 40 K and then over 50 K— under high pressure. But what really set Chu thinking were the tantalizing indications of a 70-K effect in several of his samples of La-Ba-Cu oxide. It was time for more bold thinking, but first there were a few clerical details to clear up.

Just before Christmas he heard from the efficient *Physical Review Letters* office, and as he expected, his December 15 submission to *PRL* had been accepted. The editors insisted on a few minor changes in an effort to tone down what would seem to some readers an improbable result. Thus the original title "Superconductivity above 40 K . . ." was altered to "Evidence for superconductivity above 40 K . . . ," and similar changes were incorporated in the body of the paper. The Houston contribution was scheduled to appear in the January 26 issue, the fourth weekly issue of the year, just six weeks after submission.

Brief delays to Chu's paper, presumably caused by the intervening Christmas and New Year's holidays and the need for minor revisions, were enough to allow the *PRL* editors to include in the same issue a second 2-1-4 paper, an important contribution on the strontium variant by Robert J. Cava and three coworkers at Bell Laboratories. Cava's group recorded 36-K superconductivity in the previously unreported La-Sr-Cu oxide, and they confirmed superconductivity by performing the sensitive Meissner effect magnetic measurements. Submitted a full two weeks after Chu's work, the Bell Labs paper whizzed through the publication process in less than a month. It was reviewed and accepted in a matter of days.

Many *PRL* readers saw the Houston and Bell Labs papers as complementary—not competing—studies, but Chu and his collaborators complained about what appeared to be the preferential treatment given to their New Jersey rivals. In any event, the delay of the Houston paper yielded one significant benefit. The Texas researchers were able to negotiate for a "Note added," a short paragraph outlining the Houston group's latest achievements. In that note Chu credited Wu and his Alabama colleagues with the synthesis of a strontium-bearing 2-1-4 oxide that became superconducting at 42 K. And he reported on the first tentative signs of the 70-K effect that had been seen several times in the previous weeks. Even if those remarkable results were not reproducible, Chu wanted to establish his priority. Furthermore, his observation might alert other investigators to similar phenomena in the 2-1-4 La-Ba-Cu oxides. But Paul Chu was convinced that the ultimate answer to high T_c was not going to be found in the 2-1-4 compounds. Only "dirty" samples, with a portion of phases different from the ideal 2-1-4 type, showed the transient 70-K effect. If there was a 70-K superconductor it wasn't 2-1-4. Something else was in those samples.

The original Bednorz and Müller composition had three different metal elements: lanthanum, barium, and copper. Many researchers had concentrated on substituting metals like strontium or calcium for barium, the largest element of the three and the one that might seem to alter the properties most. Copper, the smallest metal, had a peculiar combination of bonding environments in the 2-1-4 structure—environments not easily duplicated by any other metal. Furthermore, Art Freeman told Chu, the electronic structure of the copper seemed crucial to superconductivity. So lanthanum was the obvious next element to alter.

With the high-pressure data as a guide, Chu selected three

possible substitution elements. First he considered lanthanum's look-alike neighbors in the periodic table, a series of fourteen elements collectively called the "rare earths." Chu identified lutecium (Lu) and ytterbium (Yb) as likely rare-earth candidates. He also targeted the obscure element yttrium (Y), a metal slightly smaller than lanthanum. The appropriate chemicals—Lu_2O_3, Yb_2O_3 and Y_2O_3—were ordered by Chu and Wu during the first week of 1987. The search for higher T_c in new oxide systems was designated as the next project for the Alabama group under Maw-Kuen Wu, who returned to Huntsville with Jim Ashburn on January 4.

Chu and Wu had agreed to stick with the 2-1-4 composition by always mixing one unit of copper atoms for every two units of larger metal elements. Three Alabama physicists— Wu, Ashburn, and Taiwan-born graduate student Chuan-Jue Torng—prepared the two-to-one mixtures with many different compositions. The original Bednorz-Müller recipe of lanthanum and barium was altered in dozens of ways. They substituted different proportions of rare earths for lanthanum along with calcium or strontium or lead for barium. Throughout the month samples were ground, baked, cooled, annealed, and tested for electrical conductivity at low pressure. Gradually, by trial and error, they were able to raise T_c. The results seemed exceptionally promising, with frequent hints of resistance drops and other anomalies at temperatures from 50 K up to an astonishing 100 K. Acting as an extension of the Houston lab, the Alabama researchers kept in close touch with Chu, advising him almost daily of their progress and heeding his advice.

Upon learning of the startling new results, with the possibility of 100-K superconductors, Houston's Dean of Sciences Roy Weinstein urged Chu to file a patent on his latest work. Though unfamiliar with the world of patents and priority, Chu agreed to prepare something on the basis of the university's usual fifty-fifty split of profits from such a venture. Of course there was no guarantee that such a patent, if granted, would be worth anything. Chu could not be sure that he was the first to see these high-T_c effects, nor was it certain that they had any commercial applications, but it was certainly worth a try. On January 9, just before heading back to Washington, Chu quickly drafted his first patent application, claiming priority for a wide range of compositions, including the Y-Ba-Cu oxide system, and mentioning perovskite-related structures. Unfamiliar with legalese, Chu needed more than a little coaching by the University of Houston patent attorney. He dropped off the rough draft at the lawyer's office on his way to the airport. Chu's draft was polished during a marathon weekend phone conversation between Washington and Houston on January 10 and the final approved version of the patent was filed January 12, more than a month before any other research group knew of the discovery.

Under normal circumstances, with a more ordinary discovery, the 100-K preliminary results might have been put in manuscript form and dashed off to *PRL* or some other rapid publication. But Chu's lawyer said no. It was essential, he argued, to establish convincing priority over every other competitor. Secrecy and time were the best way.

By the second week of January evidence for the near-100-K effect was growing. A Houston-made sample, supposedly of the 2-1-4, La-Ba-Cu oxide composition rich in barium, was observed to have an odd pinkish color on the outside, but was black on the inside. Pei-Herng Hor made conductivity measurements on that single, odd specimen and found it to be insulating, but magnetic measurements revealed a large Meissner effect starting at 100 K. What was different about that sample? Was it improperly prepared? Was the ratio of elements different? Had it cooled at a different rate or in a different atmosphere? Chu instructed his lab team to "go home and think about it." But by the next day the sample had somehow altered. The effect was gone.

In the hopes of trying a greater number of potential superconducting compositions Chu augmented his overworked team of graduate students and postdocs with three undergraduates. They were assigned the exacting but tedious tasks of weighing chemicals, mixing and grinding powders, and overseeing the baking and cooling operations. But after three weeks of trying, nothing seemed to work. Perhaps they hadn't ground their starting materials fine enough, or maybe they mixed the chemicals in the wrong ratios. Whatever the reason, Chu reluctantly told Ru-Ling Meng, who had a special knack for the tricky synthesis, that she would have to get back to the grind.

In the midst of the growing excitement in Houston and Huntsville, Paul Chu was intrigued to read a short report in the January 17 issue of China's *People's Daily*. A week earlier, on January 11, Chu had telephoned Beijing to talk to Zhong-Xian Zhao about the Chinese superconductor research. Dr. Zhao, a former visiting researcher at Chu's lab, had become the director of the Beijing superconductor effort. Speaking in scientific Chinese, a delightful language with English words like "meters," "yttrium," and "high pressure" sandwiched in between Chinese tonalities, Chu asked Zhao if the 70-K observation—the observation mentioned in the December 27 issue of the *People's Daily*—was real. Zhao reluctantly acknowledged that the effect was "not reproducible." And in the January 17 news report the Beijing Institute of Physics group reported that while they were "working very hard on the new superconducting material," the 70-K effect was "not stable."

However, the Chinese claimed that they had exceeded 40 K in a mixed oxide of lanthanum, barium, strontium, and copper. Once again Chinese results seemed familiar. For a second time a Houston result appeared a few days later as a Beijing discovery in the Chinese newspaper. What was the real story? Was it just a coincidence that the Chinese results paralleled his own? Chu couldn't believe that anyone in the Houston lab group would betray his trust. Not even Ya-Qin Wang, a visiting research associate from Shanghai's Metallurgical Institute in the People's Republic of China, would do such a thing. But Chu did call a group meeting and he insisted that no information about the Houston research was to be revealed to anyone.

The Houston and Alabama researchers continued methodically to produce potential superconducting compounds. Time and again they glimpsed oddities in the electrical resistance suggestive of superconductivity. Transient effects near 100 K were becoming commonplace. Gradually, as evidence became stronger, the Alabama team narrowed its search to a promising composition in the yttrium-barium-copper oxide system—

approximately $Y_{1.2}Ba_{0.8}CuO_4$. That composition, the team predicted, would have the highest T_c in the 2-1-4 system. So confident were Wu and his colleagues of success that they abandoned the tedious and expensive liquid helium refrigerant needed to achieve extreme cold and made their preliminary measurements in an open bottle of cheap, 77-K liquid nitrogen.

At 3:00 P.M. on the afternoon of January 29 Maw-Kuen Wu called Paul Chu with an update. The Alabama team had prepared what they hoped would be the best sample thus far. Like the original La-Ba-Cu superconductor, the new Y-Ba-Cu mixture was fine-grained and black, but unlike the other 2-1-4 samples, the Alabama specimen possessed a distinctive greenish cast. A faculty meeting forced Wu to postpone the resistance measurement until 5:00, but Ashburn and Torng made sure everything was ready.

At 5:00 P.M. the measurement was made. Resistivity plummeted at 93 Kelvin. It was no transient effect, and Chu was notified immediately. All of the Alabama and Houston workers were exultant. It was the discovery of a lifetime.

Early on the morning of January 30—the Chinese New Year—Wu flew with the historic sample to Houston for a repeat demonstration of the effect. Given the all too familiar unpredictability of many of their high-T_c samples, there was no guarantee that the results of the previous day could be reproduced. Nerves were on edge as the tiny green-black cut prism was lowered once again into the liquid nitrogen. Once again the yttrium compound did not disappoint. The superconductivity was real.

They all knew the discovery could change the world. It was not just another perfect conductor of electricity. This superconductor worked above 77 K, the temperature of cheap, easy-to-handle liquid nitrogen: 77 K is like the sound barrier or the four-minute mile. It is *the* technological and psychological barrier against which all things cold are gauged. Below 77 K any phenomenon, no matter how remarkable, is an esoteric curiosity with few practical uses. But anyone can buy liquid nitrogen. Above 77 K there are almost no limitations to a material's applications, because almost anything can be refrigerated easily to that temperature. Paul Chu's team had discovered a material that broke the barrier. It could transform superconductivity from an oddity to a day-to-day reality.

The yttrium-barium-copper oxide was the stuff of science fiction. Chu knew that the invention of an easy-to-use superconductor would inevitably lead to a variety of futuristic inventions. The implications for society were staggering, but until this breakthrough scientists and engineers could only dream of such a compound. Now Paul Chu could afford to do a little dreaming of his own.

Cheap power was an obvious application. Superconducting cables, encased in copper and cooled with a continuous insulating jacket of liquid nitrogen, could revolutionize the electrical power industry. Power plants could be situated a thousand miles from population centers, without the substantial energy loss inherent in copper wire power transmission. Vast arrays of solar panels in Arizona could light the entire West Coast; nuclear reactors could be built in remote sites far from population centers; hydroelectric and geothermal power could be tapped efficiently in any corner of the continent. And that's not all superconductors could do. Once the power was generated, superconducting storage rings could hold the electrical energy indefinitely, without loss, ready for periods of high demand.

Computer design and construction was another industry ripe for applications of superconductivity. Superconductors had already found limited use in large, fast computers as ultrafast, supersensitive switching components called Josephson junctions. These special switches relied on a thin layer of superconducting material—a "thin film." Liquid-helium-cooled superconducting supercomputers, though costly to operate, are smaller and faster than conventional machines, which have to incorporate safeguards against the overheating inherent in conventional integrated circuits. Liquid-nitrogen-cooled superconducting supercomputers that relied more extensively on Josephson circuitry could be small, cheap to operate, and faster than anything in existence.

Magnet technology could also benefit immeasurably from the new superconductors. Electromagnets work because any electric current traveling in a loop produces a magnetic field. The superconducting magnets are vastly more efficient than conventional electromagnets because once a current starts flowing in a loop of superconducting material it goes on flowing forever, maintaining the magnetic field. Liquid-helium-cooled superconducting magnets, though limited by their expensive refrigeration systems, are only about the size of a basketball and can produce powerful fields of 5 tesla—about a million times the strength of the earth's own magnetic field —with little expenditure of energy. (A comparable conventional electromagnet with copper coils would be gigantic, requiring millions of gallons of cooling water and a prohibitive expenditure of electrical energy.) Affordable liquid-nitrogen-cooled superconducting magnets could thus usher in a new world of powerful miniature superconducting motors, near-frictionless magnetically suspended cars and maglev trains, magnetic propulsion systems, and myriads of other applications as yet unimagined.

And Paul Chu undoubtedly dreamed of fame and of wealth, not so much with hope as with concern for what they might mean to his life and work. He had not become a scientist to become a household name or to become rich. Of course, money could be used to fill his laboratory with the best equipment and support many gifted young scientists. But fame and fortune could also divert his attention from the science that he loved.

Chapter 4.

The Element of Truth

February 1987

For the time being dreams would have to wait. There was work to be done. The Houston and Huntsville teams were mobilized to tackle the Y-Ba-Cu oxide system. Wu and Ashburn flew back to Alabama on February 1 to continue their yttrium synthesis experiments, while Ru-Ling Meng, Pei-Herng Hor, and Ya-Qin Wang began to make additional specimens with different baking temperatures and cooling rates. They agreed that they should stick to the basic composition: $Y_{1.2}Ba_{0.8}CuO_4$. On February 4, through a gradual process of empirical fiddling, they had raised the onset temperature (T_{co}) to 98 K with zero measurable resistance T_{cl} at 94 K.

Pei-Herng Hor and Peter Huang attempted the low-temperature electrical resistance measurements in the presence of a strong magnetic field. Every superconductor displays a drop in T_c with increasing magnetic field. If the drop is too large, then many applications—particularly in electromagnets and motors—will be impossible. But the new yttrium sample underwent only a slight drop in T_c, even under extremely powerful magnetic fields greater than 5 tesla. That insensitivity to magnetic fields, like the high T_c, was also a record for a superconductor.

Hor and Huang next attempted measurements of the high-pressure effect. They hoped to see an increase in T_c with pressure for the new yttrium sample at least comparable to the effect seen in the lanthanum-barium-copper oxide. Two sets of resistance measurements were made, one at about 8400 atmospheres and the other at 19,000 atmospheres pressure. Nothing happened. The superconducting transition was unaffected by pressure. In one sense it was a null result, but it was one more piece to the puzzle.

As his colleagues ran the crucial experiments, Chu sought help from other quarters. The Houston/Alabama effort had already accomplished three of the four steps in identifying the new superconductor. They had repeatedly synthesized the sample, demonstrated zero resistance, and made preliminary observations of the Meissner effect. But Chu wanted to be thorough. He wanted to confirm the Meissner effect and he had to isolate the unknown superconducting phase. He contacted his friend Chao-Yuan Huang at Los Alamos National Laboratory, where more sensitive Meissner measurements could be undertaken. Samples were sent to the New Mexico lab in early February for study as soon as the equipment was available.

Then Chu considered the problem of phase identification. The new sample was clearly not a single, pure 2-1-4 compound. Though prepared in exactly the right ratio of metals to yield a 2-1-4 material, the resulting sample was an intergrowth of at least two different fine-grained substances, one of them opaque and colored black, the other translucent and colored green. The two phases could not have the same composition, so it followed that neither had a composition that matched the starting 2-1-4 element ratio. How could the unknown sample be characterized? Once again the first choice was x-ray powder diffraction.

The most efficient path would have been to give a specimen to Ken Forster and Simon Moss, who had the expertise to produce and analyze complex x-ray patterns. But Paul Chu was hesitant to let any of the precious sample pass out of his control. During the previous few weeks the superconductor team had obtained evidence that there was an industrial spy working at the Houston department. Documents were seen; phone conversations overheard. Chu and his coworkers were loathe to widen the circle of researchers with access to the superconductor secret.

So, instead of going to the crystallographers, he asked Ru-Ling Meng to do the x-ray work. Though not trained in crys-

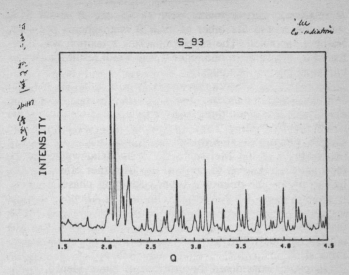

FIGURE 1. *Ken Forster at the University of Houston prepared this x-ray powder pattern of Paul Chu's superconducting sample (Chu's signature appears in Chinese characters along the left margin). The intricate graph includes dozens of lines, many of them overlapping, from two different crystal structures. The pattern is further complicated by systematic errors that were introduced to hinder unauthorized analysis. Geophysical Laboratory scientists were sent this doctored chart by mistake.*

tallography, she had often used powder diffraction to identify synthetic materials. She ground a small chip of the green-black superconductor, smeared the powder on a glass slide, and inserted the slide into the hub of the old General Electric powder diffractometer in the Materials Research Laboratory. X-rays bathed the sample with an intense stream of radiation as a detector scanned through a 50-degree arc of diffraction angles. Diffraction effects appeared automatically as peaks and valleys on a four-foot-long strip chart.

The resulting pattern was a bewildering maze of almost fifty lines. There was no obvious correspondence to any of the well-known x-ray patterns in standard reference sources. Furthermore, the twenty-five-year-old diffractometer did not have the resolution to separate many of the closely-spaced x-ray peaks; thus, even if they could identify the unknown phases, they couldn't properly match up all the features in the pattern. Chu reluctantly concluded that he needed outside help to solve the x-ray puzzle. Once again he asked for Ken Forster's help.

Forster had already heard rumors about the new superconductor and was delighted to help. Chu gave him a small piece of the precious yttrium-bearing sample, though he did not reveal the powder's composition. That information was not critical to Forster's preliminary study. By Monday, February 2, Ken Forster completed the new pattern. The data plot was beautifully prepared and ready for publication.

Two things were immediately obvious: first, there were at least two phases; second, neither of the phases was 2-1-4. It was also obvious that days of work might be needed to sort out the dozens of diffraction lines. Chu felt that there wasn't time; detailed analysis would have to wait. He decided to publish the powder pattern anyway, with Ken Forster added as a coauthor on the historic paper. After all, the knowledge that the new superconductor *wasn't* 2-1-4 was an important discovery.

Simon Moss refused. The new powder data came from *his* lab, he argued, and no one was going to publish his data without proper analysis. Moss demanded to know the superconductor composition and he insisted on time to do a complete "indexing," that is, provide an explanation for the origin of each peak on the pattern. Anything less was sloppy grandstanding—not physics, he argued.

It was Chu's turn to refuse. Rather than accede to such demands he would eliminate the figure. Chu threw down the powder pattern, harsh words were spoken, and a great rift was created between the two fourth-floor physics labs.

Ken Forster was caught in the middle. Forster's thesis advisor, Simon Moss, refused to allow further collaboration with Chu of any kind, and left Forster with the fear of failure if he disobeyed. Though he desperately wanted to work on the superconductor problem, Forster was told by Moss that it was time to get on with his thesis research. Chu demanded that Forster return all the samples and that he erase all but a couple of the powder pattern data files, which were stored in computer memory. The graduate student also decided on his own initiative to doctor the remaining files with systematic errors so that no one else could decipher their hidden meaning. Forster included a systematic angle error that subtly altered all the spacings between atomic planes; the x-ray patterns still looked correct, but they were useless without the angular correction. And so for Ken Forster the opportunity to participate in one of the century's great discoveries—a discovery happening just a few feet down the hall—slipped away.

And for the time being, without the help of the Houston-based crystallographers, Chu and his colleagues were stuck. They would have to find another way to identify the superconductor.

Amid the frenzy of experimentation there was the pragmatic consideration of patents and papers. It was imperative, once clear evidence of the high T_c was in hand, to establish priority. The University of Houston patent lawyer stressed that the patent work should be completed and filed first, before any journal submissions. Paul Chu detailed preparation procedures, compositions, properties, and the evidence for the unprecedented high-temperature effect and a "continuation" or addendum to the first superconductor patent was sent off.

After the patent came the publications. On February 5, Paul sent two papers to *Physical Review Letters*. The first, a concise seven-page manuscript, described the synthesis in a few cryptic statements and then documented the dramatic resistance measurements. The title left nothing to doubt: "Superconductivity at 93 K in a new mixed-phase Y-Ba-Cu-O compound system at ambient pressure." The three University of Alabama scientists, Wu, Ashburn, and Torng, were listed first, while six Houston physicists followed. Paul Chu insisted that his name appear last, in spite of his central role.

The paper's introductory paragraph was riveting:

The search for high-temperature superconductivity and novel superconducting mechanisms is one of the most challenging tasks of condensed-matter physicists and materials scientists. To obtain a superconducting state reaching beyond the technological and psychological temperature barrier of 77 K, the liquid-nitrogen boiling point, will be one of the greatest triumphs of scientific endeavor of this kind. According to our studies, we would like to point out the possible attainment of a superconducting state with an onset temperature higher than 100 K. . . .

It was a blockbuster of an opening. But near the end of the manuscript were two key statements that made clear there was much left to do:

Preliminary x-ray powder diffraction patterns show the existence of multiple phases uncharacteristic of the K_2NiF_4 structure.

And, just before the acknowledgments:

All present suggestions [regarding the superconducting mechanism] are considered to be tentative at best, especially in the absence of detailed structural information about the phases in the Y-Ba-Cu-O samples.

The paper about room-pressure results was to be followed immediately by the second Houston/Alabama offering, "High-pressure study of the new Y-Ba-Cu-O superconducting compound system." (Ken Forster's name was added to the author list as a sort of consolation prize for his hard, but futile, x-ray work.) The significant result was summarized in the brief abstract:

The pressure effect on the superconducting state above 77 K in the new Y-Ba-Cu-O compound system has been determined. In strong contrast to what is observed in the La-Ba-Cu-O and La-Sr-Cu-O systems, pressure has only a slight effect on the superconducting transition temperature.

Paul Chu, who had alerted *Physical Review Letters* to hold space for these historic contributions, gave the manuscripts to his secretary for typing. As the time approached for Chu to head back to Washington to confront his long-neglected National Science Foundation desk, she would prepare the typescripts, copy them, and send them off for overnight delivery.

* * *

Chu knew the risks. The journal-mandated review of his two articles might lead to premature, selective distribution of the crucial composition information. Confidential peer review is a mainstay of scientific publishing. Every journal instructs manuscript reviewers in the ethics of confidentiality: no details of a paper are to be discussed with colleagues; no result is to be duplicated or tested before publication; no benefit is to be gained from reviewing a paper. But those ideals are, in practice, difficult to maintain. Reviewers are experts who have devoted their careers to a scientific specialty. Any major advance is of immediate professional concern to the reviewer, especially if prizes and patents are at stake.

So Chu tried to negotiate with the acting editor of *Physical Review Letters*, Myron Strongin, for special treatment. First he lobbied for automatic acceptance, without any reviews. Out of the question, Strongin insisted: every paper has to go through peer review. Then Chu proposed to submit the manuscripts with asterisks in place of the crucial chemical information; the correct formula would be supplied just before typesetting. Again Strongin refused: the paper had to be submitted as it was to appear in the journal.

Chu wanted the papers published in *PRL* because it was fast and prestigious. Strongin wanted *PRL* to publish the papers because they represented a historic breakthrough. With such incentives a compromise was finally reached, though neither Chu nor Strongin was particularly happy about it. Reviewers are normally unknown to authors, but in this extraordinary case author and editor together agreed on two knowledgeable and trustworthy reviewers, whose identities have been kept confidential. The typescripts were sent by Chu overnight to the two reviewers and to the *PRL* office simultaneously. The manuscripts arrived on Friday, February 6, and within four days both papers had been accepted and were officially "in press."

The Houston physicist took some additional precautions. Access to the two typescripts was restricted. After copies of the manuscripts were sent to *PRL* and the reviewers, only one copy of each was kept, locked in the Houston files. Coauthors did not receive copies, nor were any preprints mailed out. No one, other than Paul Chu, *PRL* editors, and the two reviewers, was to know the secret information.

Did Paul Chu take other precautions? Two sets of errors appeared in the original typescripts of Chu's papers—errors that made it impossible to duplicate the superconductor discovery. First was the widely publicized substitution of the symbol Yb for Y. Every place in those two *PRL* papers where the symbol Y, for the element yttrium, was supposed to appear, the symbol Yb, for the element ytterbium, appeared instead. Two dozen times the error was repeated. The element names were never spelled out, so there was absolutely no way for an unsuspecting reader of the typescripts to catch the error.

There was another systematic error as well. The numerical coefficients in the chemical formula were given incorrectly. Evidently the coefficient "4" was substituted for a "1" (another plausible typo). No one would ever be able to reproduce the Alabama and Houston syntheses. No one could have guessed the correct formula, $Y_{1.2}Ba_{0.8}CuO_4$, from the error-riddled review manuscript.

Within a few days of the manuscripts' submission rumors of the new "Yb superconductor" abounded. The *wrong* secret was out, and at least a few scientists may have been diverted from more promising research. Does it matter? Should it?

The answer depends on one's viewpoint. Some scientists argue that if a few researchers acted on rumors and followed the false Yb lead for a couple of weeks, it was of no consequence. They should have waited for Chu's formal publication. If others misused privileged access to the Alabama-Houston manuscripts, the typographical errors served them right. They betrayed a trust. And now those scientists appear twice foolish. First, they based their research on a misprint. Second, in the ultimate, galling irony, some Yb mixtures also form a superconductor! By mid-March workers in Japan at the University of Tokyo and at Tohoku University, and in the United States at IBM and Bell Communications Research, had independently demonstrated that a Yb-Ba-Cu oxide is a superconductor above 90 K.

On the other hand, chagrined scientists blame Chu for their folly. Chu cost them time and he cost them money (pure ytterbium oxide isn't cheap). Perhaps they got what they deserved, but the episode caused feelings of resentment that are still being felt months later. Researchers at Alabama and Houston and editors at the *PRL* offices just wish the controversy would go away.

But it won't. More than any other episode associated with high T_c, the Yb-for-Y mix-up has been seized by rival scientists and the popular press as a symbol of the frantic race to discover and describe economically viable superconductors. The existence of the error, and people's response to it, epitomize the dark side of the superconductor race.

Several mysteries have framed the Yb-Y story, and few clear-cut answers have emerged. The first question, and the one foremost in many observers' minds, is how such a "typographical error" could have been made in the first place. Was it a conscious error?

Many say yes, and cite a troubling litany of evidence to make their point. The typo, they note, was extremely clever —hardly a random typing error. The metals Y and Yb, which have almost identical sizes and chemical properties, proxy for each other in nature. Furthermore, Chu had attempted to make a Yb-Ba-Cu superconductor and had failed; he must have believed that Yb would fail while Y worked. Half a dozen scientists from around the world called Chu in mid-February to see if the Yb rumor was true, but Chu responded non-committally. Some say that by not dispelling the Yb story, Chu crossed the fine line between pleading ignorance and practicing deception.

The fact of the second, little-known error in coefficients might also be explained as a typographical error, but the key to Chu's paper was the discovery of a new superconductor. It seems implausible that the typescript would be so sloppily prepared. It is entirely plausible that Chu tried to protect his discovery.

Under most circumstances Chu would have signed a letter of transmittal for the two papers when they were sent to the *PRL* office. Such letters always include the title. One source claims that a letter signed by Chu with the Yb formula exists in *PRL* files.

And Paul Chu did not call the *PRL* office to correct the

obvious errors until the last possible day—February 18, just hours before he knew the paper was to go to press. It is inconceivable, critics claim, that no one in Chu's group checked the typescripts, that no one caught the erroneous formula, before that final day.

Most disturbing of all are statements by one of Paul Chu's physics department associates, who claims that Chu showed him the erroneous typescript. The physicist discovered that yttrium was the secret element quite by accident during the first week of February. When he casually mentioned the yttrium composition, Paul Chu was noticeably upset. Soon thereafter Chu called the scientist into his office and showed him a typescript of the high-pressure paper on Y-Ba-Cu-O. The title to which Paul Chu pointed had Yb instead of Y.

If one assumes that Paul Chu deliberately falsified his formula, the question has to be asked: Was it justifiable? The majority of superconductor researchers (even several of the *PRL* editors) reluctantly say yes. It might have been easier and simpler to replace the mystery element Y—or even the entire chemical formula—with a dash or asterisk and a footnote saying that the identity of the element or formula would be revealed after acceptance of the papers. But *PRL* policy would not allow such a modified manuscript. At the time Chu was besieged by telephone calls and letters from people trying to get an edge on the competition. The world was breathing down his neck. The deliberate typos would be a simple and effective way to protect his turf. And, assuming the review process was kept confidential, no one would have cared. That the Yb information leaked so fast, some claim, fully justifies the deception. Chu had skillfully outmaneuvered his rivals.

This interpretation, which condones Chu's assumed behavior, is reasonable for a technological discovery with such great economic consequence, where patents and secrecy are vital. And there is no question that the superconductor was of tremendous technological importance. But more than a few scientists argue against that analysis, which denies a basic tenet of scientific research: it is a breach of scientific ethics to knowingly distribute false information, or even to stand idly by as it is distributed by others. Truth is paramount. Knowledge is sacred. Chu submitted his manuscript to a *scientific* journal, and he was bound to abide by scientific ethics. Thus, some scientists conclude that if Paul Chu consciously planted misprints in his *PRL* submissions then his only defense can be faulty judgment—a temporary indiscretion caused by the extraordinary tension and pressure of those frantic days.

There is, of course, another possible scenario. Paul Chu was under tremendous pressure at the time, and unconscious errors could have been made. The two handwritten manuscripts were dropped off at the departmental word-processing center just down the hall from Chu's fourth-floor office. It is conceivable that he left instructions for manuscript preparation and mailing as he rushed to catch his plane to Washington. The Houston typist may have misread Chu's scrawl, interpreting Y as Yb and 1 as 4, and sent off the internally consistent, but consistently wrong, manuscripts to the *PRL* offices. The same secretary typed the covering letter and applied Chu's signature, a practice common enough in offices. No one at Houston saw the submitted version until much later.

When Chu did return to Houston ten days later it was to a mad, circuslike atmosphere. All of his attention was directed toward an upcoming ordeal. On the morning of Monday, February 16, Chu had to present his findings before an august assembly of the University's Board of Regents and news media at a press conference that gained international attention. That Monday afternoon was fully occupied by a laboratory open house for dozens of reporters and university officials. The single copies of the two typescripts, which had been locked away for almost two weeks, were not examined until late the next day, and the corrections were hurriedly called in on the following day. Many of Paul Chu's friends would like to believe that version of the story.

As for Paul Chu, he now refuses to comment on the episode, but he promises that the truth will come out in time.

Whatever the source of the errors, the entire hassle should have ended there, as a simple confidential matter to be tidied up between author and editor before the papers ever appeared in print. But it didn't end there, and that raises a second enigma associated with the Yb-Y controversy: How did the Yb rumor ever get started? The manuscripts were mailed on Thursday, February 5, and they arrived at the reviewers' and *PRL* offices on Friday, February 6. As early as that weekend rumors of a Yb superconductor were circulating in Europe according to one Brookhaven scientist who was traveling at the time. It is difficult for me to believe, as some maintain, that Chu leaked the wrong composition himself. He made every effort to expedite the publication. Furthermore, a deliberate leak would inevitably come back to haunt him as injured parties pointed the finger. So where was the leak?

Many people have faulted the anonymous *PRL* reviewers who, being superconductor specialists and presumably working on the new oxide compounds at the time, had the most to gain from the privileged information. But with Yb rumors circulating on the weekend of February 7–8, it seems unlikely that they were the source. Reviewers had only just obtained the papers on Friday. Furthermore, Chu, who knows (but will not reveal) the identities of the two reviewers, has stated his faith in their integrity.

What about the *Physical Review Letters* office? Several Brookhaven scientists held joint appointments at *PRL*. Brookhaven is a mecca for physicists from around the country and a breeding ground for rumors. But it seems implausible that any *PRL* editor would have breached the author-journal confidence, especially given the special status of Chu's secret recipe.

But many other people might have seen one of the erroneous typescripts, either by accident or design. Did someone outside the Houston group examine the manuscripts in the word-processing center? Did a *PRL* employee casually mention the important result to a friend? Did a reviewer inadvertently leave a typescript where a colleague could see? As Myron Strongin, acting editor of *PRL*, has pointed out, at least twenty-five people might have perused either paper once they were written, and anyone could have conceivably leaked the news. It is even plausible that the Yb rumor, fueled by Chu's neutral responses, arose as an ironic coincidence at a time when superconductor rumors were a dime a dozen.

Espionage is another possibility, and if there was ever a subject worthy of scientific espionage, this new superconductor was it. Many scientists were aware of Paul Chu's ongoing research. Chu freely admitted that his next paper was going to *PRL*. And the *PRL* offices are potentially easy pickings.

All submissions to *PRL* are recorded on a central computer. Authors, title, receipt date, and review status are listed for each paper under the manuscript reference number. The *PRL* editors acknowledge that the in-house system has no sophisticated protection and any experienced hacker with telephone and computer console could figure out how to log on. If Chu's original titles, each with the Yb misprint, were added to the computer files on the receipt date, then by the weekend of February 7–8 they could have been seen by almost anyone with the requisite skill and determination.

Even more likely, a knowledgeable and persuasive phone caller could have learned the exact titles of Chu's papers just by asking a *PRL* secretary for the information. One has only to call the *PRL* office and identify oneself as an author to ascertain the current status of a manuscript.

With so many possibilities it seems likely that the source of the Yb rumor will never be identified.

In retrospect, no one seems particularly surprised that the Yb rumor got out. Far more puzzling is the apparent *lack* of any yttrium rumors after February 16. Chu corrected the Yb error by February 18, after which the revised titles and Y-bearing formula were common knowledge at the *PRL* office. Myron Strongin immediately called both reviewers of Chu's two papers to correct the formula. The key element yttrium was even inadvertently specified in newspaper articles in Houston on February 16 and 17. But no one seems to have picked up on the correct yttrium formula until just prior to release of the *PRL* issue in late February. If the wrong data leaked so quickly, how could the correct information remain a secret for more than a week *after* it was published in a major newspaper?

A third question, which has been virtually ignored by other commentators, is the identity of the individual who revealed the entire Yb-Y story to the press and scientific community. Even if there were rumors about Yb, and even if some people did pursue the wrong composition, the only people who knew *for sure* what was in Chu's original typescript were Paul Chu, the editorial staff at *Physical Review Letters*, and the two anonymous reviewers selected by *PRL*. By his own admission, it was Santa Barbara physicist Per Bak, on the Board of Editors of *PRL* at the time, who told on Chu.

Per Bak spent much of the winter of 1986–87 at Brookhaven National Laboratory. He worked both as a researcher at
Brookhaven and as one of several temporary editors for *PRL*
while regular editor Gene Wells was on leave. While at the
government lab Bak also acted as advisor to Brookhaven Postdoctoral Fellow Kurt Wiesenfeld, a physicist who is married
to a free-lance reporter, Karla Jennings. Bak was invited for
dinner at the Wiesenfeld-Jennings home on about February
20, a time when Jennings was preparing a story on *PRL* for a
local Long Island newspaper, the *Three Village Herald*. She
had originally planned to write an innocuous general piece on
the little-known Long Island physics periodical, but as Bak
talked that evening and in subsequent phone interviews, Jennings listened.

She quickly learned that superconductivity was big news,
much bigger than the daily routine of *PRL*. Per Bak described
the flood of articles on the hot new topic, he told of the potential profits and Nobel prizes, he spoke of the rivalries, and
he used the story of Paul Chu's formula as a colorful example.
Per Bak provided the scoop. Karla Jennings wrote:

> [*PRL*] has received more than two dozen papers on super-
> conductivity since January, as researchers try to claim
> whatever credit they can get. The Houston researchers,
> said Bak, even threatened a lawsuit if their most recent
> paper's contents were leaked to the public before the Hous-
> ton group held a press conference. In fact, they initially
> gave the wrong formula for their compound, changing it
> just before the article went to press. Bak said he believes
> they did so in order to insure against leaks.

Karla Jennings's article, the first to report the story of Chu's
erroneous formula, appeared in the February 25 issue of the
Three Village Herald, five days before Chu's articles officially appeared in print!

The error in Chu's coefficients was never widely known
and for a time the Yb-Y story was merely an underground
rumor, whispered in corridors of the physics community. But
eventually the story hit the news media and the popular press
had a field day. An embarrassing feature, "Yb or not Yb? That
is the question," appeared in the May 8 issue of *Science*. Staff
reporter Gina Kolata picked up the story as a rumor, and Paul

Chu inexplicably confirmed the details. Other widely read versions appeared in the *New York Times Magazine* and in a *San Francisco Chronicle* gossip column ("Physicists are almost always taught to look for the Y of things").

Per Bak committed an indiscretion—a violation of the confidence shared between author and editor. The leak of the Yb composition and Bak's subsequent revelation were an embarrassment to the temporary *PRL* staff at a time when they were under tremendous pressure to expedite an unprecedented number of superconductor contributions. The episode also left a bitter taste in the mouths of Chu and his coworkers, who were blamed for errors that should never have been made public in the first place.

And so it was that an error in handwriting—or in judgment—by Paul Chu, an information leak in an intrinsically porous scientific process, and unthinking comments by a part-time editor combined to fuel the biggest controversy of the superconductor story. All facets of this bizarre episode are not yet known. Paul Chu, *PRL* employees, and even a secretary in Houston hint of sensitive facts and confidential letters that must, for a time, remain undisclosed. As one *PRL* staffer claims, "Just wait until twenty years from now, when the full story comes out."

Before the Houston/Alabama discovery, nobody had paid too much attention to Paul Chu. But by the second week of February, as the superconductor story began to spread, he experienced the first taste of what was to come. Everyone wanted to know his secret.

One of Paul Chu's Washington-based friends, Harold Weinstock of the Air Force Office of Scientific Research, recognized possible military applications of the new high-T_c material. Weinstock asked Chu to meet with staff members at the Washington office of the Strategic Defense Initiative, more commonly referred to as "Star Wars." The SDI building was only a couple of blocks from NSF headquarters, and Star Wars money might represent a large new source of superconductor research funding. Chu had already spent some time pondering the military implications of his invention, for whenever power can be harnessed, it can be used for destruction. The new superconductor had the potential for storing, and perhaps unleashing, vast energy. It could lead to ultrasensitive infrared detectors, ultralight computers, and power storage devices. And in space, where the temperature in the shade is only a few degrees Kelvin, the compounds would need no special refrigeration. Superconductors might be the key to Star Wars.

On Friday, February 13, Chu met Weinstock for a quick lunch, after which they headed to the early afternoon appointment. The meeting was held in the office of Dr. James Ionsen, chief of SDI's Innovative Science and Technology Center. Also in attendance, at Ionsen's request, were David Nelson and Robert Pohanka, scientists at the Office of Naval Research, and Ionsen's assistant, Dwight Duston. The meeting began cordially, as Chu produced a vial of the greenish black superconducting compound and passed it around the room.

"What is it?" Ionsen asked.

"I can't tell you. We're not revealing the composition yet," Chu replied. They would have to wait two weeks, he said.

Reports differ regarding the unpleasantness that followed, and Chu prefers not to comment on the deterioration of the meeting, but according to some officials in the Defense Department angry words were exchanged. Ionsen wanted the formula, and Chu refused to divulge it. Ionsen pointed out that the Houston laboratory was supported by NSF money—government money. He asked whether Chu was obligated to release information on this government-sponsored research. Chu, fully aware of NSF policies, correctly insisted that he was under no such obligation. Ionsen invoked the names of NSF Director Erich Bloch and President Reagan's science advisor William Graham—would Chu tell them? Chu said no. Would he tell President Reagan? Again Chu steadfastly refused to give in.

"I'm very upset," Paul countered. "You're harassing me, and I will not be harassed!"

Having realized that he would secure no details from Chu, Ionsen replied, "You're wasting my time," and he walked out of his office, leaving the other five scientists to carry on without him. Paul Chu is still upset by what he feels was a needlessly harsh and confrontational experience. The ramifications of the meeting are still unclear, but one point seems certain—Chu is unlikely to request or accept any funding from SDI.

* * *

On Valentine's Day, after almost two hectic weeks catching up with NSF paperwork in Washington, D.C., Paul Chu was finally able to return to Houston for the three-day Washington's Birthday weekend. Unfortunately, it was not a weekend to relax. For the next month Chu knew he would be totally immersed in his work, without time for family and home life. And it would get worse—he was about to meet the press.

The world heard Paul Chu's news on Monday, February 16, 1987, at a news conference in Houston. The announcement was orchestrated by university administrators to coincide with the annual Board of Regents' meeting, which began that morning at 8:30. The Houston professor, along with a surprisingly large cadre of television and newspaper reporters, was on ice until the official business was done. Then, shortly after 11:00 A.M., Paul Chu faced the microphones. Following the advice of the university lawyer, he did not divulge the secret of the yttrium composition, nor did he give any hints about techniques of synthesis. Those details, Chu explained, would have to wait until the March 2 publication of his papers in *Physical Review Letters*. Instead he told of the excitement he felt at the Houston/Alabama discovery of superconductivity at 93 K and he told of its implications for society.

As he scanned the crowd he recognized the familiar face of a well-dressed Chinese acquaintance, Dr. Wen-Pu Du, a science attaché based at the People's Republic consulate in Houston. As a scientist, trained in particle physics but well versed in superconductivity, Du was evidently assigned the task of keeping tabs on Chu's work, and perhaps monitoring the activities of his compatriots at the Houston facility. Chu wondered how long Du had been in Houston, and how much contact he had had with members of Chu's physics group? Had Du acted as a conduit of Houston results to Beijing?

The banner front-page headline in the Monday evening *Houston Chronicle* proclaimed "Discovery May Earn Billions, Nobel for UH." It was unabashed hyperbole but it wasn't really distortion. The Houston announcement was the stuff of science fiction, and readers would be captivated by the news.

The reporter used phrases like "[Houston researchers] have grasped the Holy Grail of low-temperature physics," and "the basic patents stand to gain license fees amounting to billions of dollars"—a striking contrast to Chu's formal academic prose style.

But as Paul Chu flipped to the article's continuation on page five, he was horrified to see, in print, the information that had been so carefully withheld. Roy Weinstein, the university's dean for natural science and mathematics, had injudiciously told the *Chronicle*'s reporter Carlos Byars about the secret elements. There it was in black and white. After relating the original La-Ba-Cu oxide composition of Bednorz and Müller, the article revealed "Chu is now working with strontium instead of barium or yttrium instead of lanthanum." Given the original La-Ba-Cu oxide mix of the IBM group, the newspaper thus revealed that Chu was working on two systems: La-Sr-Cu and Y-Ba-Cu oxides.

There was nothing that could be done. But Chu did call Weinstein and asked him to modify any future interviews. He was assured by Weinstein that only the Houston newspaper was given the supposedly secret details. Paul Chu hoped that the Houston paper, with its limited national circulation, would not be seen by rival scientists. Of course, it was likely that the Beijing group would know. After all, Wen-Pu Du at the consulate was in Houston following the progress of Chu and his coworkers. But there was hope that American newspapers in other cities might not pick up the yttrium news. Time would tell.

The following day Chu's anxiety was compounded as he saw Carlos Byars's follow-up story in the *Chronicle*. Once again the reporter named yttrium as a key component of the 90-K compound. Again Chu hoped that the article, which was on page ten in the middle of the first section, would not be noticed by other superconductor groups. But it was in print, nevertheless.

Shortly after the Houston announcement Paul Chu returned to Washington. First there were anonymous phone calls at his National Science Foundation office, asking to know his whereabouts. Later, another caller asked Chu to confirm his American citizenship. He was advised by colleagues of "routine" security checks. Chu also met with Presidential Science Advisor William Graham and other government officials, who recognized the potential impact of his discovery.

In the midst of all the hassles and publicity a scientific problem associated with the superconductor continued to nag the Houston team. They had learned to produce high-T_c superconductors in the Y-Ba-Cu-O system routinely, but they were not making progress identifying the superconducting phase. For more than two weeks they had been frustrated. And without the help of Simon Moss and Ken Forster they were unable to decipher the complex x-ray powder patterns. Chu feared he had waited too long to act, and he decided to call his old friend Dave Mao at the Geophysical Laboratory in Washington, D.C. Mao was used to working with rocks and other multiphase samples, and he might be able to help.

The yttrium story finally broke during the last week of February. Most of the world, with the possible exception of a Chinese attaché, had missed the story in the *Houston Chronicle*. But the world didn't miss the *People's Daily*. On Wednesday, February 25, the Chinese paper announced the discovery by the Beijing group of 90-K superconductivity in yttrium-barium-copper oxide.

Several research groups in Japan, where the Beijing results were quickly seen, immediately began work on the new system, and one group at the University of Tokyo even claimed to have independently discovered the yttrium compound five days earlier, on February 20.

Scientists at Bell Communications Research (Bellcore), the divestiture-spawned spinoff of Bell Labs, heard of the Chinese newspaper article the day it was published. They were able to confirm the effect almost immediately, for as early as January 3—eight weeks before—Bellcore chemist Jean-Marie Tarascon had synthesized five samples of Y-Ba-Cu oxides. The samples had never been tested for the superconducting effect, but it took only a few hours to check that two of the five were superconducting. The Bellcore manuscript confirming Chu's results was rushed by automobile from New Jersey to *PRL*'s Long Island offices on Friday the 27th. The Bellcore driver didn't make it by the 5:00 P.M. closing, but he managed to attract the attention of a late worker and was thus able to get the February 27 "received" date stamped anyway, making the paper one of the fastest scientific studies in history.

Those scientists who did not hear about the Chinese newspaper report were not at much of a disadvantage. The March 2 issue of *PRL* was available on Long Island as early as February 25, and most East Coast laboratories received their copies on Thursday or Friday.

Paul Chu, realizing that the yttrium story was about to break and eager to gain credit for his discovery, also publicized the results. On Thursday, February 26, at 4:00 P.M., Chu gave a lecture at the University of California, Santa Barbara, in Broida Hall, the home of the physics department. The room was packed, as scientists from up and down the California coast converged on the beautiful ocean-side campus for the memorable presentation. It was there that Paul Chu first announced the secret element, yttrium. The rapt audience included Paul Grant from IBM's Almaden Research Center, where the results were duplicated on Monday after a frantic weekend of work. In addition, Chu sent out preprints of the articles to ten major superconducting laboratories by overnight express on Thursday, so that they would all arrive by Friday morning the 27th. Yttrium was no longer a secret. Before the weekend was over physicists at Bell Labs, Argonne, Northwestern, and a dozen other research teams had begun work on the yttrium compound. In a burst of activity most groups were able to duplicate the Alabama/Houston results within a few days.

Paul Chu had enjoyed four weeks in which to study the Y-Ba-Cu superconductor without competition. But there were still vital secrets to unlock. No one knew the identity of the new superconducting material. No one knew its chemical composition or its crystal structure. And the first group to solve the puzzle would have an invaluable head start in producing even better superconductors—compounds that could be worth billions of dollars. The race was on!

Part II.

WASHINGTON, D.C.

Prologue

The Geophysical Laboratory, one of five departments of the Carnegie Institution of Washington, is located on a four-acre hilltop site in the northwestern quadrant of the capital city. With an eighty-year history of research on the physics and chemistry of earth materials, the Lab supports about fifteen regular staff scientists and perhaps twice that number of predoctoral and postdoctoral fellows, research associates, and other visiting investigators each year. All of the scientists are primarily engaged in basic research, without specific economic applications as a principal motivating objective.

Our research on minerals, rocks, volcanoes, and the earth's deep interior was far removed from the concerns of those physicists who studied matter at ultracold conditions, so it was not surprising that most of us at the Laboratory were only marginally aware of the historic developments in high-temperature superconductivity. In mid-February 1987 we had other projects and other goals. We could not have anticipated Paul Chu's revelations, nor were we prepared for the extraordinary changes that the new superconductors would make in our research goals and in our private lives. We were about to embark on a scientific adventure during which the highs and lows of laboratory research—the aggravation of equipment failures and false leads, the tension of interlaboratory rivalries, the joy of discovery, and the satisfaction of sharing those discoveries with others—were condensed into a memorable four-week pursuit of knowledge.

Chapter 1.

Experiments

Thursday, 19 February 1987

I arrived at the Lab at 8:30 A.M. to find the two reviews of my hydrogen paper where I had thrown them in disgust the previous afternoon. The manuscript, submitted to *Physical Review Letters* in early December, had been rejected. I was incensed. What was perhaps my best work in a decade had been rejected as "not of sufficient interest" to the physics community. No one likes rejection, but in this case it was more than just wounded pride that fueled my anger. Knowing the paper was of interest not only to the physics community but to other scientists as well, I had no choice but to challenge the decision and resubmit the manuscript. There were experiments to monitor and the usual paperwork, but I resolved to have a new version of the study completed and ready for overnight delivery by the end of the day.

I thought back to the paper's beginnings three months before. The experiment had begun as a lark in mid-November when Dave Mao, for perhaps the third time in as many years, asked me about x-raying hydrogen. Though christened Ho-Kwang Mao, he was known to everyone as Dave for no reason except that most Americans were unable to pronounce his given name. For more than a dozen years Dave's claim to fame had been that he was the junior colleague of Peter Bell, a highly visible and well-respected member of the geophysics community. Bell 'n' Mao, Mao 'n' Bell—the names were so inextricably linked that no one, not even close associates at the Geophysical Laboratory, was ever really sure who contributed what to their scores of original professional publications on lunar rocks, high-pressure phenomena, and the rocks and minerals that occur deep within the earth. What was certain was that Dave, the quieter and more reclusive partner, underwent a dramatic change following Peter Bell's sudden departure from the Lab in the summer of 1986. Almost overnight Dave Mao became a leader of our active, if somewhat fragmented effort to study materials at high pressure. Under his gentle and persuasive guidance the work in the various spectroscopic and crystallographic labs, as well as ongoing high-pressure activities at the Brookhaven National Lab on Long Island, developed a focus that it had not previously enjoyed.

So this time I had listened with renewed interest to Dave's hydrogen proposal. For years the Lab had been conducting high-pressure studies of elements such as iron, gold, and silicon in an effort to understand the forces that dominate the earth's interior. All elements are important, but we considered gaseous elements such as neon, argon, and hydrogen to be especially intriguing. All gases crystallize at high pressure to form "solidified gases," an oxymoron that seems to have stuck. Randomly careening gas atoms or molecules, when confined at tens of thousands of atmospheres of pressure, adopt the repetitive atomic patterns characteristic of everyday solids. For most solidified gases these structures had been determined by x-ray diffraction—an experimental technique based on the interactions of a beam of x-rays with electrons in a crystal. But hydrogen, unlike most other gaseous elements, has too few electrons for routine x-ray diffraction study. So no one knew for sure the high-pressure structure of hydrogen.

The technical problems of x-raying hydrogen were formidable, but the potential rewards were great. Hydrogen, the simplest of all elements, is also arguably the most important. Hydrogen is Element Number One, the first entry in the periodic table; a knowledge of simple elements must precede understanding of more complex materials. Solid hydrogen is also one of nature's most abundant minerals, forming vast volumes—many times the total mass of the Earth—in the giant planets Jupiter, Saturn, Uranus, and Neptune. We must know the structure of solid hydrogen if we are to understand our solar system.

Given the scientific incentive, Dave argued that we should try the experiment that many believed to be impossible. He had thought long and hard about the hydrogen diffraction problem and concluded that he might have a solution. He had engineered a pressure device ideally suited to preserve the largest possible hydrogen crystal at the incredible pressure—more than 50,000 atmospheres—required to regiment the highly volatile hydrogen molecules.

Even with this experimental improvement I thought the experiment a waste of time. It was by no means certain that a single crystal of hydrogen would form, and even if it did I was convinced that the diffraction of x-rays would be too weak to detect. But Dave's enthusiasm was infectious and I agreed to try. After all, that was one of the beauties of the Carnegie Institution which, as we like to say, buys a scientist's time and gives it back, to use however the scientist wishes. That is an unusual philosophy even in academic institutions, because most scientists obtain part of their salaries from limited grant funds that usually support "safe," guaranteed productive, science. But we owed our freedom to the foresight and generosity of steel mogul Andrew Carnegie. "Uncle Andy" had benefited from scientific breakthroughs in iron processing so he decided to give something back to scientific research. Carnegie appointed a blue-ribbon board of in-internationally known intellectuals to oversee the initial disposition of the small fraction of his immense philanthropic budget that was to become the Carnegie Institution of Washington. Those experts chose to focus on aspects of botany, geology, zoology, and astronomy—choices reflected eighty years later in the Institution's ongoing studies of plants, rocks, genes, and stars.

On Saturday morning, November 22, I met Dave in my office. We walked down the corridor to the high-pressure lab where we would attempt the difficult and potentially dangerous gas-loading operation. Dave Mao was in a good humor, typical of those days when he faced an experimental challenge.

At forty-three, Dave's appearance was almost the same as it had been when I first came to the Lab a decade before—the same shaggy black hair, the same slender build, the same ready smile. Born in Taiwan to one of Chiang Kai-Chek's generals, Dave had come to the United States for graduate school and has been here ever since. He is fluent in both English and Chinese, a great advantage in dealing with the constant flow of visitors and research associates from China and Taiwan. But today, as was usual, he didn't talk much. There was hard work to do.

Our task was well defined but far from straightforward. We had to prepare a pressure chamber of diamond and steel, load that chamber with hydrogen, and seal it up. The diamond cell was the heart of the experiment. Two gem-quality, brilliant-cut diamonds were used as anvils to squeeze the sample. Between the diamonds was placed a thin stainless steel sheet, the "gasket," with a tiny hole centered over the diamond anvil faces. When pressed together, the diamond cell forms a tiny cylindrical sample chamber with curved steel walls—only as wide as the steel sheet was thick—and flat diamond ends. The beauty of the diamond cell lies not only in its simplicity, but also in its versatility. Diamonds provide the best-known window for studying materials at high pressure because, in addition to possessing great strength, they are transparent to many types of radiation including visible light, ultraviolet light, x-rays, and gamma rays. Scientists use such radiation to study the structure and properties of matter.

Our first task was to make sure that the sample chamber was well centered and secure. I say "we," but Dave Mao was the master of diamond cell design and operation; my principal contribution to the delicate gasket alignment was to keep quiet and out of his way. By noon the 2-inch-diameter cell was ready to go. The next critical step would be to load hydrogen by placing the open diamond cell into a much larger steel pressure vessel, filling the vessel with hydrogen at a pressure of more than 2000 atmospheres, and sealing the diamond cell remotely by a complex motorized gear arrangement. More than an hour of assembly was required, followed by careful checking of the high-pressure fittings with inert argon gas. We couldn't afford to make mistakes with hydrogen, which is explosive.

Soon we were ready for the most dangerous step, hydrogen pressurization. It was at this stage two years before that Dave had almost lost his life. The 10-inch-diameter steel pressure vessel, weakened by the wedgelike force of hydrogen molecules, had exploded with a deafening blast and the force of a bomb. A 20-pound chunk of angular steel had whizzed past his head and left a ½-inch-deep gash in an adjacent sheet of protective armor. Though unscathed, Dave was unable to hear for more than an hour.

Since then we had learned that special crack-resistant steels are essential in high-pressure hydrogen work and a new, safer vessel was constructed. But lingering thoughts of mutilation and death haunt the pressure lab on gas-loading days.

Despite the new system's safety features we ran into trouble almost immediately. An automatic digital gas gauge, which allowed us to supervise the operation in the security of an adjacent room, sprang a leak at 1200 atmospheres. We were forced to use a backup dial gauge mounted above the pressure vessel. We quickly lowered the gas pressure, switched gauges, and began the process anew. A video camera with a ten-foot cable monitored the gas gauge, allowing us to stand in relative safety just outside the lab's doorway. We crouched behind a TV console in the hallway.

We started the hydrogen pumps again, and watched the pressure dial slowly ascend—500, 1000, 1500, 2000 atmospheres—on up to our target of 2400 times room pressure. If the bomb were to blow, now would be the time. Unconsciously we held our breath as the motorized gear assembly tightened the hydrogen-filled cell, sealing the unstable sample between the diamonds. The danger wasn't over, but there was only one step to go: release of the excess pressurized hydrogen through a vent for safe dissipation in the atmosphere. But something went wrong again. The remote valve wouldn't open; the hydrogen bomb was still ticking. Calmly Dave Mao strode into the pressure room and turned the valve by hand, an act of bravery that he performed much as you and I would walk into a room and flick on a light.

At last the loading was complete. It had taken us six nerve-wracking hours to create a microscopic crystal of hydrogen. And now the sample was mine.

What began as a lark soon became an obsession. For days I ran experiment after experiment, searching for the telltale diffraction, never really sure if I'd ever see a thing. X-ray photographs, the normal first step in a diffraction experiment, revealed nothing at all. A more sensitive, computer-controlled x-ray diffractometer was pushed to its limit, scanning hundreds of hours for any hint of hydrogen.

It is easy to keep working on a project like this one if you are guaranteed results, but in this case I had no such reassurance. Still, having come so far, I doggedly continued. I was possessed by the experiment, driven to succeed when rational analysis told me I should fail. For days I ignored responsibilities to family and other Lab business as I sat at the computer terminal, manipulating the diffractometer that I hoped would latch onto the elusive hydrogen signal.

The breakthrough came at last as a faint x-ray signal that just barely appeared above the level of background radiation. At first I thought it only my imagination, but by adjustments to the x-ray source and detector the peak was enhanced. Given that one beautiful diffraction maximum it was a simple matter to locate other x-ray peaks. Within minutes I had solved the puzzle of solid hydrogen's structure—a simple and elegant pattern in which each hydrogen molecule is surrounded by a symmetrical pattern of twelve other molecules. Some theorists had predicted that this structure, seen in many metals, would also accommodate hydrogen at pressure. Our experiment had provided the proof. With a sense of urgency I dashed off a ten-page manuscript and sent it overnight to *Physical Review Letters*, a logical choice for publicizing fundamental new physical data on Element Number One.

Now, two months later, the preliminary verdict was in. Rejected! It was not so much the worry of getting published that bothered me. We all knew the manuscript would be accepted somewhere. But the delays had created a problem with our follow-up experiments. Our initial success prompted a flurry of additional hydrogen x-ray measurements at much higher pressure—work that effectively superseded my effort. It was essential to get the initial paper accepted so the second could be submitted. With a renewed sense of frustration I studied the two reviews.

Physical Review Letters relies on two referees for every manuscript submitted. Like most scientific journals, *PRL* keeps the reviewers' identities confidential, so we couldn't be sure who had judged our work. What was certain was that we got a split decision, and a split decision means rejection.

Referee A was laudatory, recommending publication and asking only for an illustration to show a typical hydrogen diffraction peak shape. That was an easy detail to clear up. All the diffraction data was stored as digital computer files, and Larry Finger, fellow crystallographer and systems manager of the Lab's in-house computer network, was the one to create the needed figure.

I found him seated behind his desk amid a jumble of partially written manuscripts, unread scientific journals, and thick computer printouts. A tall, medium-built man in his mid-forties, Larry has a full, thick beard and a mass of curly gray hair that retains only vestiges of its original brown. Despite the cold February weather, he wore his usual short-sleeved shirt, a reminder of his hardy childhood on a Minnesota farm.

I proceeded to show him the two reviews. As coauthor on the hydrogen work, Larry was visibly annoyed, but his actions were constructive. He immediately attacked his computer terminal's keyboard, calling up the hydrogen diffraction file and converting it to a format for plotting. Within five minutes I was holding two neatly printed peak profiles, one of the strongest hydrogen diffraction and the other of a relatively weak maxima. Referee A was right—the inclusion of illustrations would improve the paper.

Next it was time to tackle the review of Referee B. In addition to numerous minor suggested changes, all easy to incorporate or rebut, he or she had two damning objections. First, B claimed we were "unbelievably ignorant of the past two decades of theoretical work on solid hydrogen." Second, even if we did greatly expand our discussion of theory, B felt our "neat little experiment" was not of sufficiently general interest to merit *PRL* publication. Both comments smacked of a desk-bound theorist who had never run an experiment in his or her life.

B did have a point about one thing though. I *was* totally ignorant of recent hydrogen theory. And I wasn't about to learn it all in a day, either. So, taking the path of least resistance, I decided to ask Rus Hemley, coworker on the second round of hydrogen diffraction experiments at Brookhaven, to "fix" the theory section in exchange for being listed as a co-author on the paper. Rus, fresh from a doctorate in chemistry at Harvard, is our newest staff member. Though only thirty-two, he had already combined a masterful experimental technique with broad-ranging theoretical ability. Most of us would probably feel threatened by his expertise were it not for his quiet nature and easygoing willingness to share his knowledge. Though we sometimes call him "the Golden Boy," the nickname is rooted less in jealousy than admiration.

Rus was working intently at a computer terminal when I entered his spectroscopic lab. True to form, he didn't seem disturbed by the interruption. He stood up when he saw me, stretched his arms back and grinned. "Guess it's time for another cup of coffee, huh?"

Of medium height, Rus Hemley is compactly built. He has dark blond hair that he keeps fairly long, possibly in an effort to hide his fast-receding hairline. He wears glasses and, like most of the noncrystallographic staff, he's clean-shaven. As he poured coffee he asked for the latest news from the x-ray lab.

"The hydrogen paper was rejected by *PRL*," I said, coming right to the point. "One reviewer liked it, the other said it wasn't of general interest. You know *PRL*—one bad review and you're shot down."

"Those guys are so inconsistent," Rus sympathized. "They'll accept something incomplete on high-pressure iodine and then reject high-pressure hydrogen. You should definitely resubmit and ask for another opinion." Having just struggled to get a paper on silicon glass into the journal, he spoke from experience. "Will you have to change anything?"

I showed Hemley the negative comments and made him the proposition. He gladly agreed to beef up the theory in exchange for coauthorship, and he went straight to work. I left him with a copy of the manuscript and walked back to my office, composing a letter of resubmittal in my mind.

The letter had to be carefully planned. On the one hand, we strongly disagreed with B's opinion, and we had to rebut the conclusion about relevance. On the other hand, B was presumably an expert on hydrogen, and we couldn't just dismiss the comments as uninformed. I adopted a formal tone and resorted to the stock phrase: "We have carefully considered the two reviewers' comments and have revised the manuscript accordingly. In particular..." After a brief synopsis of the alterations, I concluded by restating our position on the importance of the work and requested another set of reviews. By midafternoon the letter was typed and I was completing a Federal Express form when the phone rang.

It was Margee, calling from home. She sounded cheerful but purposeful, and I suspected she needed help or advice because she rarely called just to chat. She asked how the day was going but I wasn't in the mood to tell her about the hydrogen hassles. Somehow I dislike talking about my failures.

She got to the point quickly. "Is there any chance you could come home a little early? I have string quartets tonight, but we have to help Elizabeth get started on her science project. She's chosen evaporation as her subject and she's pushing to compare plain water with cinnamon water. I don't know how to handle it. Is that really an OK experiment?"

Evaporation of cinnamon water is not a problem I'd thought much about, but it probably seemed quite interesting to eight-year-old Elizabeth. Though academically precocious, her principal goal in life at the moment seemed to be acquiring holes in her ears. She could be quite determined and if she had decided to do the experiment, she wasn't likely to be dissuaded.

"It's not the experiment I would have thought of," I said, "but it certainly is an experiment. She's asked a question: 'Does cinnamon water evaporate the same way as plain water?' and she's going to find out the answer. I guess it's OK." I thought a moment and added, "Maybe we could convince her to try a third solution like salt water as well."

We talked for a minute or two about the size of the glasses and the quantities of cinnamon and salt, and it struck me how similar Elizabeth's science project was to our own day-to-day work. Ask a question, perhaps guess the answer, devise an experiment, run the experiment, and interpret the result. It always sounds so simple when you put it that way. It would be fun helping Elizabeth and I promised Margee I'd be home early to act as science advisor.

* * *

There was always so much work to do at the Lab, but there was nothing that wouldn't keep for another day or two. The past months of hydrogen study, with long hours and lost weekends, plane trips to Brookhaven, and other professional obligations had made me at times almost a stranger in my own home. I didn't want the temporary pattern to become a habit.

I leaned back in my chair and gazed at Margee's portrait, taken about twenty years before. She still had the beautiful dark blond hair, though not quite so long and with a well-concealed gray strand or two. She still had the slender figure, delicate features, and thoughtful sea-blue eyes—haunting Botticelli eyes, I always thought. We'd shared a lot of good years since we met in Mr. Bowler's high school algebra class.

In spite of busy schedules, we had always made time for each other. I was determined not to let that change, especially now that we had established an enjoyable and productive writing collaboration to complement our family life. Together, we even survived the tedium of proofreading and indexing. But our latest book, on brass bands, was finished in November and the next, on fire technology, was only in the proposal stage. The hydrogen experiment, coming as it did at a time of transition, demonstrated how easy it would be to drift apart, to become immersed in our separate little worlds.

Having decided to leave the Lab early, I phoned Rus and he agreed to send off the revised manuscript with my letter and mailer. There were just a few odds and ends to tidy up when Dave Mao walked into my office. As he entered he closed the door behind him, an act signaling confidentiality that we rarely employ.

I assumed he wanted to talk about hydrogen—it was obvious he wasn't in my office to engage in idle conversation—and his secrecy piqued my interest.

"How small a crystal can you measure?" he asked without preamble. "Not hydrogen—just a regular oxide."

"It all depends on what you want to measure," I told him. "What's the composition?"

"I can't tell you that," he said, and urged me to estimate a size anyway. Dave doesn't often beat around the bush, so this cryptic response surprised me.

"Well, maybe 20 microns." I'm not sure why I made that guess. Such a crystal would be too small to see with the naked eye. Never before had we worked on anything that minuscule and I think Dave knew it. But he didn't react other than to nod and depart with a quick word of thanks.

Little did I realize as I grabbed my briefcase and headed toward the car that my bold words would lead to a month of frantic days and sleepless nights. Little did I realize that all of our other scientific projects, hydrogen included, would soon seem unimportant. We were about to begin the quest of a lifetime.

Chapter 2.

Changing Directions

Friday, 20 February 1987

My morning commute to the Lab wasn't far, but it wasn't fast either. I joined three solid lanes of Connecticut Avenue traffic, bumper-to-bumper, all heading south to D.C. It's like this most mornings. I inched ahead in a dead heat with a Porsche 944 to my right.

At least my ten-year-old son Ben was along to keep me company until I dropped him off near his school. He was silent, staring out the window, thinking about things small boys think about, I suppose. His straight brown hair was only partially combed, as usual, and I noted with some alarm that he was outgrowing yet another pair of jeans and winter coat. Though more inclined toward the visual arts than science, Ben dutifully reported to the science classroom early three mornings a week. His love of animals prompted him to care for the school's menagerie of insects, frogs, lizards, and hermit crabs. Having reached his corner, he gave me a perfunctory "Thanks, Dad," slipped on his red nylon backpack, and left me to continue alone.

It was just me and ten thousand other commuters crawling to work on one of Washington's main arteries. Fortunately the tourists had not yet invaded to compound the inconvenience. As I finally arrived at the Lab's sweeping driveway I noticed the time. All in all it wasn't a bad morning—only twenty-three minutes to travel seven miles.

The eighty-year-old research facility is a solid though not

imposing building on Upton Hill, one of D.C.'s highest points. The Geophysical Laboratory, affectionately known as the "Gee Whiz Lab" by almost everyone except the forty-odd scientists who work there, has a reputation as a home for a productive, if eclectic, band of geologists, geochemists, geophysicists, and others who don't quite fit any neat category.

The quaint pseudo-Spanish stucco exterior and red tile roof of the main building conceal massive two-foot-thick solid masonry walls and a foundation firmly planted in bedrock. The architects, trying to prepare for any experimental needs, made every effort to design a vibration-proof environment. Other unusual features, including a wide central corridor, oversize windows, and fourteen-foot-high ceilings, were intended to accommodate scientific apparatus undreamed of in 1907. But as most laboratory experimental hardware became more compact—and critical climate control much more expensive—the grand old edifice was increasingly viewed as inflexible and costly by some Carnegie administrators. A modern, fully equipped laboratory building is now planned for a larger site about a mile to the north. We will all welcome an up-to-date facility, but for most of us the Geophysical Laboratory will always be that odd, historic, red-roofed building on Upton Hill.

Friday is usually a day for wrapping things up, for clearing off the desk. The day's schedule seemed easy enough. Scientists from Ohio, California, Italy, and Australia had sent postcards requesting copies of my papers on crystals at high pressure. It's always a simple matter as well as an ego boost to send off reprints, so those envelopes were addressed first thing. Other obligations would take more thought. Two manuscripts needed editing for *The American Mineralogist,* and another had to be reviewed for *Journal of Geophysical Research.* Trivial final revisions were also required for a paper that was close to acceptance in *Physics and Chemistry of Minerals.* So much for the paperwork.

The x-ray laboratory demanded little attention. Our research included high-pressure crystal structure studies on garnets, pyroxenes, and perovskites—a trio of intriguing minerals thought to occur deep within the earth. At the heart of the crystallography lab are three x-ray diffractometers, prosaically named "North Picker," "South Huber," and "Picker Number Two." All three are complex machines that measure the interactions of x-rays with crystals. The Cleveland-based Picker Instrument Company designed and manufactured our two older diffractometers in the 1960s. Picker diffractometers are built like tanks and are painted institutional green. They're slow, but they seem to run forever. Huber in West Germany built our newest diffractometer. It's faster, more streamlined, and is painted a pleasing shade of blue.

Small computers control our diffractometers. Each was programmed to collect data automatically through the weekend. What a boon computers have been. Thirty years ago crystallographers operated diffractometers manually; they worked long hours to collect the kind of data that I would find neatly recorded on my printout Monday morning.

The week was supposed to end, as usual, with an informal Friday afternoon seminar. The in-house speaker is chosen by lot the day before and the Lab supplies the beer. Tom Hoering, our resident organic chemistry guru, had been tapped to entertain us with stories of his recent struggles to study sulfur isotopes in coal and other biologically deposited rocks. Tom has cultivated a somewhat bemused view of scientists and the scientific process, so his presentations are always seasoned with witty and entertaining commentary.

In short, I probably would have had a relaxing, low-key sort of day if Paul Chu hadn't entered our lives. It was just before noon, almost time for the daily (weather permitting) volleyball game, when Dave Mao came unannounced into my office with a serious expression on his face.

For the second time in as many days he closed my office door behind him as he entered. I motioned for him to sit down and he pulled a chair close to my desk. Hesitating at first, Dave pulled a small ordinary-looking vial out of his shirt pocket and carefully handed it to me.

"Paul Chu wants us to study this material," he said.

I waited, expecting Dave to elaborate, but he seemed reluctant to say more. I racked my brain trying to remember where I had heard Paul Chu's name. I was sure he was not an earth scientist, but we also interact with materials researchers from physics, chemistry, and other fields.

"What is it?" I prompted. As I waited I peered through the

bottom of the glass and shook the contents. It was nothing more than a dull black clump of irregularly shaped grains. A few isolated dust-sized bits had broken off the main mass and littered the bottom and sides of the half-inch-tall container. It looked like a typical sample of iron oxide—not very interesting.

"It's a new high-temperature superconductor that Paul just made." Again Dave paused, apparently not willing to volunteer anything.

Superconductors! That was it. Paul Chu was one of the many material scientists working on those rather exotic curiosities. His special contributions I seemed to recall had to do with measuring superconductivity at elevated pressure. He had shown that superconductivity in some materials is enhanced by subjecting samples to high pressure. Chu studied that effect in the hopes of finding ways to improve samples at room pressure.

"You mean higher than the 30-K stuff?" I asked. We had all heard about the IBM Zurich discovery of a new class of oxide-based superconductors. Perhaps Chu had improved on this "record" high temperature.

Dave looked me in the eye. "This compound is superconducting at 90 Kelvin."

Even though I didn't immediately grasp the full significance of his assertion, that simple statement may have been the single most remarkable thing that anyone has ever said to me. 90 K! The unbelievable breakthrough had been made! The implications were obvious: a technological revolution might result from Paul Chu's discovery. And I held the stuff in my hand.

My stunned silence prompted Dave to continue. "Paul tried to identify the material with x-ray powder diffraction but the pattern is complex and doesn't match any known structure. It looks as if there may be more than one phase in the sample, but he only knows the bulk composition. We have to find out how many different phases there are and we need to determine their compositions and structures. Can you help?"

There was no way I could refuse. Paul Chu had just handed us the problem of a lifetime. He had discovered a remarkable new compound but he did not know its chemical composition or crystal structure. Our laboratory was ideally suited to determine those properties. We were being given a unique opportunity to study the most important new material to come along in decades. "This is fantastic!" I responded. "Yes, sure I'll do it. Of course I'll have to know Chu's starting composition."

It was an obvious request but one that, in this instance, had far-ranging implications. The superconductor formula was worth a fortune. Whoever had that formula had a vital head start on the competition in designing and patenting superconducting wires for power transmission, switches for computer applications, and other products of inestimable value. Paul knew that the Geophysical Lab, with no expertise in materials processing, was not about to engage in patent-oriented research. But we could inadvertently reveal—or purposefully sell—the vital information to one of Chu's rivals. So, before disclosing details of his superconducting formula, Paul Chu had sworn Dave Mao—and anyone Dave chose to work with —to absolute secrecy. No one outside of the research team was to be told anything about the starting composition. Furthermore, as we deduced details of the number of different phases and their individual compositions and structures, the new information was to be given to Chu and no one else, especially prior to the anticipated early-March appearance of his key *Physical Review Letters* paper announcing the discovery. Considering the historic importance and financial implications of the sample, we were more than happy to agree to these understandable stipulations.

Dave Mao wrote the formula on a small slip of paper and handed it to me:

$$Y_{1.2}Ba_{0.8}CuO_4$$

The superconducting sample contained oxygen bound to three types of metal atoms. Yttrium—an obscure element retrieved as an afterthought from mines in obscure countries. Nobody ever seemed to talk much about yttrium, but that would change. Barium—a massive, prosaic element found as a trace chemical in ocean water and concentrated in sediments from ancient, dried-up seas. And copper—the everyday element of pots and pennies. Paul Chu mixed these three humdrum elements and committed alchemy. He turned them into gold.

Even better than gold, for the yellow metal is merely valuable —the superconductor is also magic. And we were to help find out how the trick worked.

Chu had given us a synthetic sample and told us its bulk chemical composition. He indicated that the specimen was a fine-grained mixture of two or more different materials, only one of which was the superconductor, and he gave us Ken Forster's x-ray powder pattern. Our job was to isolate each of the sample's distinct compounds and to determine the chemical compositions and crystal structures of each of these phases. So, with little fanfare or formality, Paul Chu set us on the quest for the superconductor. It took only a few minutes to radically alter the focus of our research for months to come.

Why did Paul Chu choose Dave Mao? As in so many human interactions, friendship was the key. The two had first met at a high-pressure meeting in the mid-1970s. Dave, fresh from graduate school at the University of Rochester, reported on his pioneering geophysical studies with diamond-anvil pressure devices. Paul's work on superconductivity at high pressure was described at the same meeting. The friendship begun at that conference was reinforced by subsequent scientific gatherings and a collaborative study of copper chloride at high pressure. Even though they usually worked on different materials, their common emphasis on pressure was a scientific thread that bound them as fast friends.

Furthermore, Dave represented a laboratory that was well qualified to study mineral-like compounds, yet had no vested interest in superconductors. Many other materials science groups could have analyzed Paul's samples, but few could have done so without some threat of information leak or other impropriety. The Geophysical Lab, which worked largely outside the superconductor and low-temperature physics mainstream, was an ideal choice.

My first response to the revelation of Chu's discovery was one of wonder and excitement, yet it is difficult in retrospect to recall my exact thoughts. It is true that I abruptly dropped other projects, though none of them was of an urgent nature. I did stay late at the Lab and worked with an intensity matched in recent years only by the hydrogen studies. But the staggering implications—scientific, economic, societal—dawned on me only gradually over the next days and weeks. I was not at first keenly aware of competition from other laboratories, nor did I contemplate spending so many frantic days and sleepless nights in pursuit of the unknown atomic structure.

But Dave Mao and I both sensed a unique opportunity in the dull greenish black crust of superconductor. We knew the task would be difficult, but we were also confident of success. And so, while Paul Chu was not there in person, his presence was keenly felt, and all of us who were privy to the superconductor secret shared in the exhilaration of its discovery and the urgency of the race.

Our problem was to take this enigmatic black sample and to learn what made it unique. Knowing the three crucial metal elements was not enough because Chu's superconductor was a mixture of at least two *different* crystalline compounds—two different "phases," each characterized by different ratios of elements. Pick up almost any rock and you'll see a variety of grains, some clear and colorless, some black or darkly colored, some milky white. Each of those different kinds of grains is a distinct mineral in which oxygen binds together with a different combination of metal elements. Clear beach-sand quartz forms from silicon and oxygen; black magnetic magnetite incorporates iron and oxygen; white feldspar arises from a complex oxide mixture of sodium, aluminum, silicon, and other elements. Together these and other oxygen-based minerals make up most rocks.

The new superconductor was a synthetic rock; though unlike any earthly rock in the specifics of its bizarre composition, it too was a mixture of different mineral-like grains. For each of these "minerals" we had to know the exact ratios of yttrium to barium to copper as well as the details of the atomic patterns formed by those elements that created the superconducting crystal mass. We had to design an experiment.

Measuring compositions and determining atomic structures are exactly the sorts of problems that every geologist is faced with when studying a newly found rock. So from our point of view the experiment really wasn't all that different from the day-to-day, routine study of any unknown sample. But there was a critical difference between a study of natural rocks and the impending investigation of the superconductor. Most rocks have rested, unchanged, for millions of years before succumbing to the geologist's hammer and their study is, correspondingly, generally accomplished at a leisurely pace of years or even decades. But Paul Chu's superconducting rock was cooked up in a laboratory just a few weeks before, and his problem had to be solved in a matter of days. For, as he was quick to emphasize, others would soon be hot on the trail.

No momentous scientific discovery can long remain a secret, especially in the open academic environment of nonprofit university research laboratories. At least ten researchers at the University of Houston and the University of Alabama participated in the superconductor discovery and its initial description, and all of them knew the fabulous consequences of their work. Editors and manuscript reviewers at *PRL*, and patent lawyers and university administrators at Houston also knew. Rumors of Chu's discovery had been circulating for weeks and the facts would soon be published for all to see. There was no time to waste.

Why such urgency? Billions of dollars were at stake. The key to understanding how the superconductor works—and thus the key to fabricating superconductors that would work at even higher temperatures—lay in knowing its exact composition and atomic structure. The first research group to solve the structure could be the first to deduce those potential improvements. And Paul Chu wanted us to be first. Though I had never before collaborated with the Houston scientist (indeed, we had never even spoken about our scientific work), our working relationship was clear. Chu had provided us with a unique opportunity to study a sample of historic importance, and we could provide Chu with the vital data he needed to improve the material. Our relationship was scientific symbiosis.

Lunch forgotten, Dave Mao and I went straight upstairs to the laboratory. Our first step was to look at the sample in a light microscope. Under magnification the black compound had a rough, granular appearance with a pronounced greenish sheen. Little else could be discerned from the aggregate so Mao and I decided to crush a 1-millimeter chip in oil between two glass slides. This procedure produced myriads of minute fragments, most of them irregularly rounded with no obvious crystal faces. The average size was no more than 5 microns (two ten-thousandths of an inch) in diameter. By observing these particles in a powerful transmitted-light microscope, we could see what appeared to be two different compounds: two distinct phases, each with its own distinctive color and shape. In order to solve the superconductor riddle we would have to discover the chemical compositions and atomic arrangements of both phases.

Most of the superconducting sample consisted of an extremely fine-grained substance that was a dark emerald green, a color characteristic of dozens of copper-bearing minerals. So intense and deep was the color that fragments had to be extremely thin for any green light to pass through. Nevertheless, this compound gave the entire black superconducting mass its greenish luster. The individual green crystals were much too small to handle with our normal techniques. So, after some discussion, we reluctantly decided to defer working on the difficult green phase. It was a decision dictated by the limitations of the sample more than by our own rational choice.

Searching the sample further for something we could work on, we focused on a few larger crystals that appeared to be black and opaque and more regular in shape than the green phase. Materials that conduct electricity are usually opaque, so this black compound, though it made up much less than half of the superconducting sample, was well worth a closer look. Several particles between 20 and 30 microns in size seemed to be the best single grains we were likely to find. We agreed to begin work on those "black-phase" crystals.

Dave Mao and I had been huddled over the microscope for more than twenty minutes and our conversation was at times quite animated as we plotted our research strategy and debated the relative merits of various black specks. Larry Finger, with his office adjacent to the crystallographic lab, came out to see what all the excitement was about. We quickly swore him to secrecy and apprised him of the situation, knowing full well his expertise would be essential in solving the structures.

"90 Kelvin! That's amazing!" He, too, was excited about the discovery. But when he looked down the microscope and saw the fine-grained sample with crystals smaller than anything we'd ever attempted to study, he just sighed. With that he left the crystal mounting and preliminary x-ray diffraction to me. However, before he returned to his office he did offer to come into the Lab on Sunday if I needed help. I told Larry that I might take him up on the offer if I ran into trouble.

Dave Mao took a portion of the material downstairs to mount for compositional studies while I began to hunt for a suitable x-ray specimen. The first step in any single-crystal experiment is to find a good crystal. Normally that means a cube-shaped fragment about 100 microns in length on each side, or about half the size of a grain of table salt. But these opaque particles were much smaller than that, averaging only 25 microns in diameter and impossible to see without the aid of a microscope. After almost an hour of searching speck by speck I found one that appeared to be both large enough to mount and fairly regular in shape. With the utmost care that tiny crystal had to be isolated from the thousands of other particles in the oil drop. I manipulated an ultrasharp needle, gently pushing and prodding the grain through oil toward the edge of the slide. Separated from the main oil drop, the tiny crystal stuck safely to the needle's pointed end, which was used to transfer the black speck to a second glass for cleaning in acetone. The slightest slip at that point—even a strong breath—could have lost the crystal forever. But fortunately the mounting operation was a success. I glued the crystal neatly to the tip of a slender, half-inch-long glass fiber, where it could be bathed in the probing beam of x-rays.

X-ray crystallography is one of the marvels of twentieth-century science. Scientists had long suspected that crystals were composed of regularly repeating sequences of atoms, though the atoms themselves were much too small to see. They even suspected that these regular planes of atoms were spaced a few billionths of an inch apart, though no one at the time knew how to measure such incredibly small dimensions. Wilhelm Roentgen's serendipitous discovery of x-rays in the summer of 1895 changed that.

Physicists soon recognized that x-rays were just another form of light. Though of much higher energy than visible light, x-rays can also be recorded on photographic film. And just as visible light diffracts off a finely striated surface such as a phonograph record or diffraction grating, so x-rays diffract off the finely spaced planes of atoms in a crystal.

When an x-ray beam strikes a crystal, x-rays are diffracted off the lattice and radiate out in different directions, forming a regular and symmetrical pattern of x-ray "reflections." For each arrangement of atoms there is a unique pattern; the pattern shows up on x-ray film as spots and in an x-ray counter as peaks. The key to unlocking the atomic structure of a crystal is to measure the intensity of these reflection patterns.

In 1912 the German scientist Max von Laue applied the elegant mathematical formalism of crystallography to the phenomenon of x-ray diffraction and showed how to predict every x-ray diffraction direction, given a known atomic arrangement. The mathematics of the diffraction process aren't trivial: x-rays don't "see" atoms the same way a medical x-ray "sees" bones. But Laue proved that an atomic arrangement leads to a distinctive diffraction pattern of x-ray intensities. Seventy-five years later x-ray diffraction measurements are still the best way to determine crystal structures. All we would have to do, then, would be to work Laue's process in reverse: measure the diffraction pattern and identify the atomic structure that would produce it.

After the minimal one-hour drying time, I placed the freshly mounted black speck on an x-ray camera. Was it a single crystal or just a useless glassy mass? Would I see any diffraction spots? I positioned the sample halfway between the x-ray-producing tube and a special, x-ray-sensitive Polaroid film cassette and began a conservative twenty-minute exposure. The first picture, developed at 3:00 that afternoon, was a success. Four weak but clearly visible white spots graced the black background of the film. There was no particular symmetry to the pattern—the spots appeared to be spaced at random. But any x-ray diffraction was better than none.

I took a second photograph after rotating the crystal 90 degrees; this time five spots were present. I was in business! With nine distinct spots on two photographs the crystal was ready to be transferred to the South Huber x-ray diffractometer. Unfortunately, the garnet experiment in progress had to be stopped, but no one was about to scoop the Lab on that research. Our priorities were clear. The new crystal was attached and centered.

The four-circle diffractometer, with its massive geared metal rings and jutting arms, is complex in appearance but simple in concept. A fixed source of x-rays irradiates the crystal, which is located at the center of four motorized arcs. Three of these arcs are used to move the crystal into any desired orientation, while the fourth arc cradles a sensitive detector that measures the intensity of diffracted x-rays. It takes only a few minutes to position the diffractometer arcs and locate diffraction effects corresponding to spots on x-ray film. An automatic peak-centering computer program is then initiated to provide more precise locations for each reflection.

I selected the strongest spot on the first film and directed the motorized arcs to the correct positions. A weak peak was there; the detector registered about 100 x-ray counts per second compared to a surrounding background of fewer than 10 x-ray counts per second. The entire centering process took perhaps fifteen minutes. Then on to a second spot and a third.

Though driven to push on with the work, I had to break away to hear Tom Hoering's talk. Everyone was expected to attend, and under ordinary circumstances it would have been a perfect end to the week. Tom is the Lab's only practitioner of old-school chemistry. In this day of black-box chemical analysis, automated chromatography, and vibrational spectroscopy (in all of which he's an expert), Tom still knows how to blow intricate vacuum-tight glassware and pour acetone from a five-liter jug into a two-ounce stopper-bottle without spilling a drop. His description of sulfur isotope work was novel and amusing, for while sulfur is an essential ingredient of life, it leads to some of the least pleasant odors we know. But I couldn't pay much attention. There was so much to do on the superconductor; so many questions remained unanswered. The immediate concern was six more x-ray spots to center. Fortunately Tom stuck close to the half-hour time limit. Then it was back to the Lab.

With several more hours of preliminary x-ray work to do I called home to report on my unexpected delay. I had anticipated a relaxed evening at home with Margee, so I had mixed feelings about staying late. She seemed to understand and decided to feed the children soon, delaying her own dinner until I made it home. She was curious to know what had come up at the Lab, and I was eager to tell her about the superconductor discovery. But Paul Chu's warnings made me hesitate. I'd tell her later, in person, when no one could accidentally overhear.

After our good-byes I rushed back to the x-ray lab. The remaining six spots were found and centered, one by one. Some were very weak, registering only about 50 x-ray counts per second on our detector. As a result it took longer than anticipated to center them all, but they were all there. By 7:30 that first evening I was ready to try a computer procedure known as "auto-indexing."

There are two major tasks in determining a crystal structure, which is composed of countless identical tiny building blocks, called "unit cells." First, the crystallographer must measure the size and shape of that building block: the unit cell might be a cube or a rectangular prism or some other six-sided form. Trillions of trillions of these boxlike units are stacked side by side, end to end, and top to bottom to form a crystal. Second, the scientist must deduce the arrangement of atoms inside each unit cell. A few simple structures have only one or two atoms per unit cell, but complex structures may have hundreds or thousands of atoms in each identical building block. This second step must follow the first.

Auto-indexing is a crystallographic shortcut for completing the first step—determining the unit cell. The procedure relies on a computer's ability to search rapidly for regular mathematical patterns. I had centered nine x-ray peaks, each of which could be represented by a line radiating from the tiny sample. The situation was something like having a sea urchin with all but nine spines missing; analysis of nine remaining spines might suggest positions of the missing ones. Similarly, there was a good chance that the computer would recognize a systematic pattern of peak positions based on the nine known x-ray peaks. The computer could then point the way to other peaks—and the unit cell. That's the way it was supposed to work, in theory anyway.

I hurriedly typed the observed diffractometer settings for all nine spots into the computer and waited expectantly for the auto-indexing calculation. The response three minutes later was a letdown:

NO SOLUTION

What was wrong? Perhaps in my haste I had typed the settings incorrectly. I quickly proofread the columns of numbers and found the error. A "34" was transposed to "43." Again I typed the numbers, this time a little more slowly, and again called the auto-index routine. Three minutes later:

NO SOLUTION

For some reason there was no simple structure that could explain all nine observed peaks. Not sure what to do, I drove the diffractometer back to the first reflection. It wasn't there! After a couple of minutes searching, the errant spot was found, almost a full degree of arc from where it had been four hours earlier.

I cursed in frustration as I realized the problem. The glue that secured the crystal had obviously not dried sufficiently after mounting and, consequently, the crystal had been shifting position, ever so slightly, throughout the afternoon's work. All of my peak centerings were invalid because all had changed as the crystal moved. Auto-indexing couldn't possibly work. There was nothing else to be done. I'd have to start all over again tomorrow morning.

The Lab phone rang as I was about to leave. It was Margee. "Hey, remember me? It's after eight o'clock and I'm sort of hungry. Are you by any chance playing with hydrogen again?"

I sighed. "No, it's not hydrogen this time. I'll tell you all about it when I get home. I'm on my way, I promise."

I kept my word and left a few minutes later. But I'd be breaking a lot of other promises in the next four weeks.

Chapter 3.

The Black Phase

Saturday, 21 February 1987

The sky was still dark at 6:30 A.M. It was too early to get up on a Saturday, but I had slept well and felt driven to get back to work. Quietly slipping out of bed so as not to disturb Margee, I threw on an old pair of jeans and a sweater and headed downstairs for a hasty breakfast. I was back at the Lab by 7:30.

Everything was as I had left it. My first task was to relocate the slowly shifting reflections that I'd seen the night before. It was a tedious, repetitive process. For each of the nine spots I had to position the crystal and x-ray detector at the approximate diffractometer settings that I'd measured the previous night, find and center the reflection manually, and then wait ten to fifteen minutes for the computer to determine the precise positions of the peak. Within two hours all nine spots had been relocated. All nine had moved slightly during the night, but the glue finally seemed to have dried, stabilizing the crystal.

I typed the new peak settings; now auto-indexing was sure to work.

The computer disagreed:

NO SOLUTION

I reviewed the possibilities. All nine of the reflections were sharp and clearly belonged to the tiny black sample. All of them were of similar modest intensity, from 50 to 150 x-ray counts per second. And, curiously, the nine reflections represented diffraction from only two different spacings of atomic planes. That fact implied that the crystal possessed a high symmetry.

Symmetrical objects appear the same in more than one orientation. A plain white piece of paper looks the same front and back; it has twofold symmetry. Many pencils have six identical sides; they have sixfold symmetry. Cubes have six identical faces, eight identical corners, and twelve identical edges, a three-dimensional symmetry called cubic.

Crystals, too, have symmetry. Crystals appear the same to x-rays in more than one orientation whenever the arrangement of atoms is identical in more than one atomic plane. In a crystal with cubic symmetry, for example, the six atomic planes that are parallel to cube faces are all the same. Therefore, the intensities of x-ray reflections from these equivalent planes are also the same. The characteristic angles between x-ray reflections from different cube faces are always 90 or 180 degrees, the same angles that are observed between the cube faces themselves.

One key to recognizing a high-symmetry crystal is identifying certain characteristic angles between pairs of reflections. Angles of 45, 60, 90, or 109.5 degrees all suggest types of crystal symmetry. So I began to check all of the angles formed by pairs of the nine centered reflections. The process was tedious. With two reflections there is only one angle; with three reflections there are just $1 + 2 = 3$ angles to calculate. But by the time I reached nine observations, there were $1 + 2 + 3 + 4 + 5 + 6 + 7 + 8$ or a total of 36 separate calculations to perform. It took the better part of an hour, sitting at the computer terminal and typing all the peak settings associated with each reflection pair. The results were not encouraging. There were only a couple of angles near 45 degrees, not enough to recognize any special symmetry. Furthermore, several of the angles were less than 10 degrees apart, unreasonably small for most structures.

X-ray reflections are like regularly spaced rays emanating from a tiny crystal; angles between adjacent rays should, in general, be greater than 15 or 20 degrees. The most likely reason for the observed smaller angles was multiple crystals. I suspected, though could not yet prove, that my single black grain was composed of several tiny crystals stuck together. The overlapping ray patterns of multiple crystals could easily have angles of less than 10 degrees between rays of two different crystals. Normally under such circumstances I would have abandoned the difficult sample and sought another, better crystal. But there was no guarantee that a better crystal could be found and time was a dominant factor. We would have to be creative. But scientific creativity is often implemented by tedious, repetitive experiments.

The only way to be sure of the number of separate interlocking crystals in the black sample was to commence a systematic search of all possible diffraction effects for the two most common atomic spacings. I set the x-ray power to maximum and began an automatic peak-searching routine, a process that would take many hours. The computer proceeded to rotate the crystal systematically into every possible orientation, while the x-ray beam and the x-ray detector were fixed at positions that would spot anything similar to the strong reflections I had already centered.

It was less than ten minutes before the first spot showed up. Within an hour there were two more. I was in trouble, for at this rate there would be more than thirty reflections to center, and more than 500 angle pairs to compute. I was more sure than ever that the sample was a multiple crystal, and there was no way to sort out which reflection belonged to which crystallite without knowing all of those 500-plus angles. It was time to call Larry Finger.

Shortly after 1:00 P.M. I reached him at home. I tried to sound good-natured as I asked him to give up part of his weekend. He seemed happy to help and he asked what needed to be done. I described the problem of the large number of reflections and my suspicions about multiple crystals.

"Do you want me to spell you centering reflections?" he asked. He knew how tedious that job could be.

"Actually, what I could really use is a computer program to calculate all of the angles between reflections," I replied. "You know, just like a string of GOSUB120 commands." I had learned that Larry was as fluent in BASIC computer language as English, and so I referred to a section of his program that ran the diffractometer system. I was hoping that he could get the computer to perform automatically the hundreds of tedious angle calculations that I had been typing in by hand.

Larry agreed without the slightest hesitation and promised to come into the Lab early Sunday afternoon. He sounded enthusiastic; writing computer programs, perhaps even more than square dancing with his wife Denise, was his love in life. And he could easily imagine how essential such a program might be in this instance. I was delighted at his response and I promised to have a list of thirty or forty reflections to try.

With the auto-search routine working smoothly and the promise of Larry's help I felt much more relaxed. I decided to have lunch and call home to check in and report my progress. I suddenly felt generous with my time and proposed that the family take an afternoon excursion. Given the rarity of such opportunities I assumed that my offer would be accepted without hesitation. But Margee was neutral in her response. Evidently she and the kids were in the middle of projects. "Why couldn't you have mentioned this last night or before you left?" she asked.

She was right, of course. It would have been nice to plan ahead. But science is unpredictable. An experiment set in motion can go almost anywhere, and scientists can't set themselves a rigid timetable. Margee knew that was a fact of our lives and I was annoyed that she seemed to have forgotten it.

"Oh, never mind. I'll try to be home for dinner." I hung up abruptly without waiting for a response. It was a stupid move, for it could only alienate her. The pressure was getting to me already.

Larry's Lab phone rang almost immediately. It was a number Margee almost never used, but she had it written down for emergencies. She was more upset than I'd expected and she wanted to know what was wrong.

I calmed down quickly, hoping she would excuse my irrationality, and tried to explain that I couldn't tell when free time was going to come up. I knew the superconductor experiment might take a long time, and with upcoming trips to Chicago and Long Island I had just wanted to do something together as a family when the rare chance came along.

She responded more calmly as well. "I'd like to do something, too, but everybody's in the middle of something. Can you free up some time tomorrow? We could plan a trip for then."

I thought about the schedule and realized that Larry was going to take over in the afternoon. It might be the best chance. "How about two o'clock tomorrow afternoon?" I sounded much like a businessman setting up an appointment. "That looks clear. Let's make it then."

I was not alone at the Lab in my superconductor studies. While I had been engaging in preliminary x-ray work Dave Mao had commenced his investigation of unknown chemical compositions of the phases that composed the greenish black sample.

Dave knew the bulk composition of the material with its 1.2:0.8:1.0 ratio of the yttrium, barium, and copper components. But there were at least two different phases; two discrete chemical compounds formed Chu's synthetic rock with its mixture of black and green mineral-like grains. We hoped that each of these "minerals" would be made of a simple ratio of the three oxides. But there was one analytical problem; the black and green phases were intimately intergrown at a scale that could only be resolved in a powerful microscope.

There was a time not many years ago when the analysis of such a microscopic mixture of grains would have been all but impossible. Each phase would have to be isolated by hand under a microscope—a Herculean task, something like separating grains of salt and pepper that have been accidentally shaken together and then ground to a fine powder. Fortunately the invention of the electron microprobe in the 1950s transformed the science of chemical analysis.

The main body of the electron microprobe is a powerful electron gun that produces an intense beam of electrons, accelerates those electrons toward the sample target, and focuses the electron beam to a tiny spot. These energetic electrons are absorbed and then reemitted by each element in the unknown sample. As a by-product of the electron interaction, each excited element produces a spray of x-rays—x-rays with distinctive, characteristic energies for each different element.

The heart of the electron microprobe is a remarkably sensitive and complex array of detectors and associated electronics that measure the energy and intensity of the millions of x-rays produced in the probe. The number of characteristic x-rays detected for each element is directly related, though in a non-linear way, to the amount of that element in the sample.

In a sense, therefore, Dave's job seemed straightforward. Just mount a bit of the oxide on a glass disk, coat the sample with an electron-conducting layer of carbon atoms, and measure the composition. But nothing about this superconductor problem was going to be easy. First, in order to get good quantitative results, the sample needed to be smooth and polished, and we had never tried to polish 5-micron grains before. Second, the number of characteristic x-rays emitted by a given weight of sample differs for every element, each of which requires some kind of standard. We had no yttrium standard to put in the probe.

The solution to those problems, and hence more precise quantitative analyses, would have to wait until our microprobe expert, Chris Hadidiacos, came in later in the week. In the meantime Dave Mao was only able to perform an ambiguous qualitative analysis.

He placed the 1-inch-diameter sample into the probe's small vacuum chamber and waited for the pump to do its work. A green light signaled all was ready. Dave manipulated several of the machine's many dials, thus focusing the electron beam to the smallest possible spot, blasting the sample's surface with electrons and precipitating the desired spectrum of x-rays.

Dave watched the results on a colorful video display. All three metal elements—yttrium, barium, and copper—produced strong signals, recorded as peaks on the video console. In the best situation, if two different phases were present in large discrete grains, there would have been only two distinct patterns, each with peak-height ratios corresponding to the element ratios of one of the two phases. But virtually every pattern Dave tried looked slightly different. In some the yttrium peak was suppressed relative to barium and copper, but no regular pattern emerged. It was obvious that the 10-micron-diameter electron probe beam was too large to resolve individual 5-micron grains of the mixed-phase superconductor. Even at that extremely small scale, the beam invariably sampled mixtures of black and green particles. It would be another few days before unambiguous composition data were available.

Sunday, 22 February 1987

Some people's lives follow the comfortable rhythm of the recurring weekend, but during the past several months of hydrogen work I had enjoyed no such luxury. With afternoon family plans set, I had to be at the Lab early. I wanted to make sure that all was in order for Larry's arrival. The auto-search routine had taken much longer than I anticipated, and it still had perhaps an hour to run. Rather than wait, however, I stopped the operation and began to center the newly found reflections. There were almost forty maxima to locate, so I knew that I'd have to settle for rapid, approximate, manual positioning rather than more precise but time-consuming automatic centering.

Slowly I plodded through the list, checking off about ten of the spots per hour. Several of them corresponded to peaks that I'd already found and centered the previous day, so I was able to include more precise values of their diffractometer settings. Slowly a pattern was emerging.

About a third of the reflections, which tended to be the more intense ones, had x-ray detector positions corresponding to periodic (repeating) planes of atoms spaced 2.7 angstroms apart. Two-thirds of the reflections were weaker and had detector settings that indicated a 1.9-angstrom spacing. I concluded that those two types of atomic planes were probably the two strongest diffraction effects from a crystal of high symmetry.

Shortly before 1:00 P.M. I had located all the maxima. I took a clean piece of paper and tabulated the sets of four angles that defined the orientation of each reflection in four neat columns. All of the stronger 2.7-angstrom peaks were listed first, while the 1.9-angstrom reflections followed. Then it was up to Larry Finger.

He arrived a few minutes after 1:00 in an enthusiastic good humor. I showed him the long list of centered reflections and pointed out the two characteristic atomic spacings of 1.9 and 2.7 angstroms. He must have realized that I'd been working hard, but he seldom offered compliments and none were forthcoming for my efforts. Instead he just went to his desk and began to create the program that would solve our problem.

Knowing my help was no longer required, I shut off the x-ray generator and said good-bye. We agreed to meet first thing in the morning to run the program that was already taking shape in his mind.

Things had been progressing very well. To the best of our knowledge no other group had learned the yttrium secret and we'd made a good start. We had already isolated a crystal of the black phase and soon we should know its unit cell. Of course there was a lot of work to do—and that fine-grained sample did leave a lot to be desired. But we were well on our way to a structure solution and I was glad for a chance to take a break.

Margee and I had decided to visit historic Gunston Hall, the Colonial American home of Virginia patriot George Mason. The afternoon was cold and overcast and smelled of wet weather but we enjoyed a relaxed tour of the house and grounds and then commenced an easy two-mile hike to a Potomac River overlook.

Ben and Elizabeth ran ahead, throwing sticks and playing tag, while we followed at a more leisurely pace. Margee slipped her arm through mine as we walked along.

"I'm glad we could do this today," she said. She paused and glanced at me briefly. It must have been obvious that my mind was back at the Lab because she immediately turned the conversation to the experiment.

"How's it going? Are you happy with your progress?"

It was difficult to say. "Well, yes and no. We haven't really gotten very far yet, but we will. It's such a tough problem—one of the hardest things I've ever worked on."

We moved on. "You know," I continued, "this superconductor thing is probably the most important problem I've ever worked on. It's an incredible opportunity."

We said nothing for a while as we proceeded past a half-frozen pond and followed the trail up a steep incline. As the path leveled out she picked up the discussion again.

"But you're sure you'll be able to figure it out?"

"Oh, we'll get it eventually. The real question is whether we'll be first."

It was going to take a tremendous effort. That much was clear. But I was determined to keep the momentum up until the end. I really wanted this one.

Monday, 23 February 1987

The precipitation had begun as a gentle white dusting late the previous afternoon, but within a few hours an intense storm of heavy, wet snow was blanketing the Washington area. By 1:00 A.M. our house, along with tens of thousands of other suburban dwellings, was without electricity. Trees and power lines across the metropolitan vicinity were toppled as the sticky, dense snow accumulated by the ton, bowing or breaking all but the sturdiest limbs.

By morning the snowfall had ceased, but streets appeared impassable with a nine-inch layering of snow punctuated by innumerable fallen trees and branches. In front of our house a neatly cleaved thirty-foot maple had missed my car by a mere three yards, and another large uprooted tree blocked all but a few feet of road farther up the street. Everything was canceled, from schools and businesses to public transportation and the United States government. It was an enforced holiday that most people welcomed, for the fresh white snow turned the landscape into a beautiful frosted fairyland. But I was obsessed with getting to the Lab. I was sure that Larry's new programs would give us the unit cell, and with that information we could begin to solve the black-phase structure in detail.

So at 8:00 A.M. on the worst day of the winter I began my foolhardy trip. The snow cover was thick but uniform and the Toyota just barely cleared its surface. I proceeded cautiously but with increasing confidence as the small car sliced steadily through the undisturbed blanket. I could even steer deftly through the narrow gaps formed by curb and fallen limbs without slowing appreciably.

Then, to my surprise and delight, I saw that the main cross street ahead had been plowed. If I could make it that far I'd be home free. There was just one more barrier to pass—the considerable wall of hard-packed snow thrown up by the snowplow that had cleared the thoroughfare. I revved the engine, accelerated from first to second gear, and learned a lesson in physics.

I made it about halfway through (or more precisely over) the pile when my forward momentum ceased. I had managed to plane the vehicle onto the barrier pile so that the car's undercarriage rested comfortably on the snow base while its wheels spun uselessly a fraction of an inch above nothing but air. I was only two feet laterally from the now-drying main road surface, but it was the unexpected vertical separation that stymied me.

There was nothing to do but grab my snow shovel—at least I had come prepared—and begin the laborious liberation process. It took an exhausting twenty minutes of excavating —chiseling, scraping, digging, and throwing the hard-packed snow and ice—before I was free. But from there on it was about the easiest commute I'd ever had, with only a handful of other cars on the snow-free main roads. I pulled into the deserted Lab parking lot by 8:40.

I ran upstairs and found to my relief that all was working as planned. The Lab was evidently in one of the few areas that had not experienced a power outage. I immediately set about recentering several of the strongest black-phase reflections and tabulating all of the observed maxima from the last two days. I knew that Larry would be in soon and I wanted to be ready for him. Actually there was no compelling reason for Larry to commute on such a hazardous morning. He had a home computer link to the Lab and could do almost as much work at home. But I knew that he would come in. He loves the challenge of a little snow, and takes great pleasure in showing up the inept Washington winter drivers with his Minnesotan's skill. True to form, he was in by 9:30 and immediately set to work showing what his neat new computer program, ANGLE, could do.

The program's input included the diffractometer settings for each of the forty-odd observed reflections that I had measured. The output was a table of all of the hundreds of angles between pairs of these reflections. Almost at once the answer was obvious to Larry and me. Dozens of pairs came up with either 60, 90, or 120 degrees, clear signs of a crystal with cubelike symmetry. We selected a group of six reflections, all of which showed these special angular relationships, and fed them into the auto-indexing routine. For the first time in days the program worked!

The unit cell was indeed cubic—the basic building block of the black-phase structure was a tiny cube 3.9 angstroms, or about 0.000000015 inches, on an edge. A reexamination of the complete set of forty observed reflections revealed the source of our previous confusion. As I suspected, we had been working with a multiple crystal. The black-phase "crystal" that I had mounted was actually a composite of at least four different grains stuck together. Each of these four grains had an identical unit cell but each was in a different orientation. It was no wonder we couldn't decipher the complex overlapping x-ray effects. But now that we knew the cubic unit cell, we were in business. We could instruct the computer to ignore all but one of the individual crystals. We identified the largest of the grains on the basis of x-ray diffraction intensity and started to improve our measurements of the unit cell by careful recentering of the strongest diffraction maxima.

Larry and I were elated at the morning's events, for we knew we were well on our way to a structure solution. Not only had we discovered the simple unit cell, but we were pretty sure we knew the basic atomic arrangement as well. There are only a few simple structures with small cubic unit cells, and each has a distinctive x-ray pattern. Chu's black phase corresponded exactly to one of them—the "perovskite" structure, an atomic pattern adopted by almost all of the previously known oxide superconductors.

Scientists have identified hundreds of compounds with the perovskite crystal structure. The first such material to be described was calcium titanate ($CaTiO_3$)—the mineral perovskite, from which the group gets its name. Other perovskites include barium titanate ($BaTiO_3$), a synthetic compound used every year in billions of dollars' worth of electronic gear, and magnesium silicate ($MgSiO_3$), a high-pressure mineral that is believed to make up more than half the volume of the solid earth. These and other simple oxide perovskites have the general chemical formula ABO_3, where *"A"* and *"B"* represent metal atoms that combine with three oxygens denoted "O_3."

The basic perovskite structure is simple and elegant (Figure 2A). The large *A* metal atom lies at the cube center, smaller *B* metal atoms occupy the cube corners, and oxygen atoms occur halfway along all cube edges. One consequence of this atomic structure is that every small metal atom is surrounded by six oxygens in an octahedral arrangement (Figure

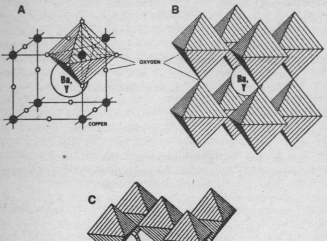

2B). Each of those octahedral units shares all six of its "corners" (i.e., the oxygen atoms) with six others, giving the perovskite structure an infinite framework of corner-sharing octahedra.

But not all perovskites are that simple. The straight rows of octahedra in the ideal cubic form can tilt in dozens of different ways to yield lower-symmetry variants. Metal atoms can be shifted off center from their ideal positions to yield commercially valuable perovskites with unusual electrical properties. A fraction of the metal or oxygen atoms may be missing, leading to a bewildering array of "nonstoichiometric" perovskites, that is, crystals whose formulae do not have simple integral ratios of atoms (for example, $ABO_{2.63}$ or $A_{0.86}BO_3$ instead of ABO_3).

Thus, while we had identified the simple cube-shaped unit characteristic of all perovskite-related structures, Larry and I knew our job had just begun. We were sure that the black perovskite was nonstoichiometric—the chemical formula couldn't possibly be ABO_3—because of simple electrostatics.

FIGURE 2. *On February 23 Larry Finger and I discovered that the superconductor has a perovskite-type structure. The simplest perovskites (Figure 2A) have a cube-shaped unit cell. Smaller "B" metal atoms (for example, copper) lie at cube corners, a larger "A" metal atom (such as barium or yttrium) resides at the cube center, and oxygens occur along each cube edge. Adjacent unit cells share atoms on edges and corners to make a continuous, three-dimensional structure. The average unit cell content is one A, one B, and three O atoms—or ABO_3. The perovskite structure can be illustrated in different ways; for example, by highlighting "octahedral" clusters of six oxygen atoms that surround each B metal atom.*

Octahedral clusters form a continuous network in perovskites (Figure 2B). Every octahedral corner represents an oxygen atom, while copper atoms are not shown in this schematic drawing.

The simple cube-shaped perovskite can be altered in a number of ways. Calcium ferrite ($CaFeO_{2.5}$), for example, has fewer oxygen atoms than simple perovskites, thus altering some of the octahedra to smaller polyhedra (Figure 2C). This perovskite has a doubled unit cell, a larger repeat unit than the simple perovskite cube. Missing oxygen atoms and a large unit cell were keys to solving the superconductor structure.

Barium, yttrium, and copper atoms usually possess ionic charges of $+2$, $+3$, and $+2$, respectively. No combination of two of these metals (corresponding to the A and B in ideal ABO_3) could yield a $+6$ charge to balance the -6 charge required for three oxygens (the O_3 in the formula). Yttrium and copper, for example, might combine to form a perovskite with the formula $YCuO_{2.5}$, but never $YCuO_3$. Electrostatic charge balance is an essential requirement for all oxides, so we knew that the ideal ABO_3 formula had to be wrong. The ratio of oxygens to metals had to be less than 3 to 2, most likely because of missing oxygens rather than extra metal atoms. We could labor for weeks trying to find the nature of those subtle deviations from the simple, ideal perovskite structure. Yet those deviations, and their effects on the electronic behavior of the black phase, hold the key to understanding high-temperature superconductivity.

For the time being there was nothing else to do but proceed with the precise unit-cell measurement. That procedure would take about twenty-four hours, after which time we could worry about solving the structure itself.

I sought out Dave Mao, hoping I could tell him the news before his scheduled departure for Brookhaven that afternoon. He would not normally have driven the considerable distance from his home in Fairfax, Virginia, to the Lab on such a snowy day, but he had much to do before the next round of hydrogen experiments in New York. As I suspected, he was in. He was pleased, though not surprised, at our findings, and in return he gave us his preliminary microprobe results. One of the phases in the original sample was significantly enriched in yttrium compared to copper and barium, whereas another seemed to be yttrium-poor. The mixture was so fine grained, and the grain boundaries so irregular, that it was impossible to be sure which phase was which or even if any of the first analyses represented the composition of a single phase. Mao's information was not of great help at this stage of our work, and I hoped that better composition data would follow.

After some more discussion of experimental strategy we agreed on one more step that should be accomplished as soon as possible. We headed upstairs to report to Charles T. Prewitt, an internationally known crystal chemist and the Geophysical Lab's new director. Charlie Prewitt had moved steadily up the scientific ladder from industrial chemist to university professor to department chairman. In July 1986 he succeeded Hatten S. Yoder, Jr., as director of the Geophysical Laboratory, only the fifth individual to hold that post in the Lab's eighty-year history. In spite of his new administrative responsibilities he maintained an active research program on such far-ranging topics as the sizes of atoms, calcium minerals at high pressure, and, in his spare time, the schooling patterns of fish. But what was foremost in my mind was the x-ray diffraction work he had done years ago at DuPont on economically important mineral-like compounds not unlike our superconductor sample. Common courtesy required that we inform the director about the abrupt change in our research program, but this was more than an obligatory call. With his knowledge and expertise in powder x-ray diffraction, he could become a valued member of the superconductor team.

We hoped that Charlie could answer a nagging question connected with the superconducting sample. We knew that we had a mixture of at least two different compounds—one black and one green. But was there a third? We might have two different types of green crystals or two different opaque phases; we'd never be sure from the single-crystal studies, alone. But Chu's powder diffraction pattern—the pattern obtained by Ken Forster in the Houston crystallographic lab—included peaks from *every* compound in the synthetic material. Our single-crystal x-ray studies revealed the atomic spacings of individual black and green crystals; Charlie could use those data to check off all the peaks that came from the black and green phases. Any leftover peaks on Chu's x-ray powder pattern would signal the presence of a third, as yet unrecognized, compound.

We found him hard at work, his jacket thrown across the back of his chair, his shirt sleeves rolled up. At fifty-three he was a slender man with graying hair and fair skin. Charlie sat at an enormous oak desk, a relic of an earlier era, as he slogged through the paperwork that was now a basic part of his job. Thanks to his hard work the other Geophysical Lab scientists were almost completely insulated from the bureaucratic chores that often accompany academic and governmental research jobs.

As soon as he saw us at the door he invited us in, but in his usual quiet manner he said nothing more, waiting for us to begin the conversation.

"Have you heard about the new developments in superconductivity?" I asked.

He nodded slightly, cuing me to continue.

"Well, Paul Chu at Houston has a 90-K superconductor and he's given us the sample to characterize." I went on, describing the multiphase nature of the sample and emphasizing Paul's understandable desire for secrecy.

Charlie listened intently while I talked, maintaining eye contact while sitting motionlessly. Only when it was clear that I had told my story did he speak.

"That's really exciting," he said with just enough inflection to establish his fascination with the unexpected developments. "What can I do to help?"

It was the response we'd hoped for. We showed Charlie the complex x-ray powder pattern that Chu had given us and the preliminary perovskite cubic cell and asked him if he might be able to analyze them.

"I think so," he replied. "I've just transferred all of my powder diffraction programs to my new computer in the next room. They're all up and running. I'll get on it right away."

We handed him all of the pertinent charts and tables and left him to his work, knowing that our chances of success had just improved considerably.

I returned to my office and reached for the telephone. Part of our deal with Paul Chu was to keep him informed of any new developments. I assumed he'd want to know of the perovskite finding.

His office phone was answered after only a single ring by an efficient-sounding secretary. I briefly identified myself and told her my business. "Dr. Chu gave us some of his new superconductor sample. He asked me to call if we got any information on it."

She told the standard white lie, "I'm not sure if he's in. Can you hold?" With a click the line became silent, with only the faintest hiss of soft static, but within a few seconds Paul Chu answered in person.

"Hello, Bob! This is Paul. It's good to hear from you. How

is everything?" Though I had never met nor spoken with him before, his warm and cheerful first-name greeting was delivered as if we were old friends. His animated voice, graced by a singsong inflection, had only a mild Chinese accent.

"Hello, Paul," I responded, adopting his familiar style. "We have some preliminary information on the black phase." I proceeded to tell him about our first x-ray experiments and the perovskite-type cube unit cell. I explained that our next step was to test for possible perovskite superstructures composed of more than one cube unit.

He made notes on our x-ray results and then asked if we had any compositional data. I outlined Dave Mao's preliminary electron microprobe findings and promised that we'd have more details by the end of the week.

Again Chu thanked the Lab for its efforts, but he added a word of caution. "My paper in *Physical Review Letters* will be in the March second issue—that means it will be out by the end of this week. We don't have much time."

I assured him that we'd do our best and thanked him again for the chance to work on the problem. Our business concluded, we exchanged good-byes.

Without putting the receiver down I broke the connection and immediately dialed home. It had been hours since I checked on the family and I was hoping that the power problems might have been resolved. No such luck. Margee and the kids had just finished a lunch of hot dogs roasted over a fire in our family room fireplace. The house was cold, there was no hot water, and the food in the freezer was thawing.

I commiserated on her loss of the amenities of modern life and told her of our success in discovering the perovskite-like structure of the black phase. Margee, however, was more interested in the practical applications of superconductors—specifically if they would prevent power outages during blizzards and thunderstorms. I had to admit that even with superconductors nature would still sometimes call the shots. Nothing we do will stop trees from falling on power lines.

Chapter 4.

Northwestern

Tuesday, 24 February 1987

Tuesday morning saw a small exodus of mineral physicists from the Geophysical Lab. A team of six scientists, including Larry Finger, Rus Hemley, Dave Mao, and Charlie Prewitt, headed to Brookhaven by car and plane to gain a last few precious days of high-energy synchrotron radiation before the National Synchrotron Light Source ceased operation for eight months of renovations. They would attempt to push the hydrogen experiments to record high pressures before shutdown on Friday night.

Part of me wanted to go with them to Long Island, and part of me wanted to stay with the superconductor in Washington, but I had to fly to Chicago instead. Several months earlier I had agreed to present the Ipatieff Lecture at Northwestern University and even though February can be a terrible time to fly to O'Hare, complex schedules left no alternative date. In spite of our commitment to Paul Chu and the obvious need for speed, all work on the superconductor had temporarily come to a halt. My main concession to superconductivity was the decision to return to D.C. immediately after the Wednesday lecture rather than joining the Long Island group as originally planned.

My flight out of National was not scheduled until 10:00 A.M. but the vagaries of D.C. traffic demanded an early departure from home and a brief stop at the Lab just to make sure all three diffractometers were operating properly.

Indeed, the Lab was running smoothly. The South Huber diffractometer had completed its unit-cell measurements on the black-phase granule sometime during the night and it sat patiently waiting for its next instructions. The output confirmed that the basic structural subunit was a cube, about 3.87 angstroms on a side, but we had to find out if more than one cube combined to form the unit cell. A multiple-cube unit cell might possess properties—including superconductivity—that weren't present in a simpler perovskite. While I was in Chicago the diffractometer could check out several possibilities. Doubling of the subcell is among the commonest of perovskite variants, so I set up the x-ray system to sample all possible reflections in a 2 by 2 by 2 cube-shaped cell. Within a couple of days I should know for sure if any crystal axis was doubled.

I mounted a magnetic tape, checked the power settings on the x-ray tube, and initiated the program that controlled a systematic collection of x-ray intensities, the next step in the structure determination. That data collection would continue for at least as long as I was away. There was nothing else to do at the Lab so I headed to the airport.

Chicago, cold and fair, had less snow on the ground than Washington. I hoped that pattern would persist for another thirty-six hours. My Northwestern host, chemistry professor Duward Shriver, was waiting near the gate.

As we drove to Evanston, Shriver briefed me on the schedule for the next two days and told me a bit about the beautiful Northwestern University campus, situated on a mile-long tract beside Lake Michigan. Following a brief tour the rest of the afternoon was taken up in informal meetings with individual faculty members. The superconductor problem was never far from my thoughts, and I found it difficult to concentrate on the visits. Only one conversation made a lasting impression.

"Conversation" is perhaps an inappropriate description of my meeting with Herman Pines, for we did not discuss our research or scientific philosophies. Rather, he told me the remarkable story of Vladimir Ipatieff, the man who had endowed the prize and lectureship that bears his name.

* * *

Ipatieff made his fortune in the chemical industry by discovering a high-pressure catalytic process that turned crude oil into gasoline. Though his initial experiments were in the realm of pure chemistry, he quickly recognized their potential applications and obtained the appropriate patent rights in the 1920s.

Ipatieff's odyssey began in turn-of-the-century tsarist Russia where he was a brilliant chemistry student. Recognizing the value of more advanced training, his family sent the young scientist to Germany where he supplemented his chemistry education and made the close personal contacts that would eventually save his life. Although the Russian Revolution of 1917 was a time of turmoil for many academics, Ipatieff did not suffer. Lenin knew that the young Soviet nation needed a solid technological base, and he championed Ipatieff as a scientific leader in the new Russia, encouraging him to maintain his European contacts, travel abroad to scientific gatherings, and even take out international patents on his chemical discoveries. Ipatieff became a highly privileged man in a society where privilege was hard won and often came at a price.

Stalin's rise to power in the 1920s marked a dramatic change in Ipatieff's fortunes. The benefits he enjoyed under Lenin were, in themselves, sufficient to make him suspect. Close friends in Russia, sensing his peril, urged him to flee. In the summer of 1930 Ipatieff and his wife made one last trip to the West, ostensibly for a meeting of German chemists—and didn't return. They had escaped just in time, for many of Ipatieff's scientific colleagues were never heard from again.

The Russian expatriate was invited to join the fledgling Northwestern University chemistry faculty in 1931. There he made his new home, set up his own company, and reaped the benefits of his pioneering research.

In the Ipatieff saga I saw a close parallel to the story of Paul Chu. Chu was a foreign-born chemist who sought new opportunities in the West. Paul's superconductor research was of a basic nature, yet it, too, led to discoveries with important implications for the energy industry. I could only hope that Paul Chu would gain the scientific and financial recognition enjoyed by Vladimir Ipatieff.

The afternoon one-on-one meetings were followed by a reception at the Department of Chemistry facilities and a pleasant dinner with a half-dozen chemists and geologists. When the evening's activities were concluded, at about 9:00, Shriver took me to my hotel, an elegant, opulent structure with plush carpets, mirrored walls, and gilt rococo decor. My room was spacious and comfortable, but I felt strange and out of place. Though fatigued, sleep seemed impossible. The things I really wanted—the superconductor sample and Margee—were a thousand miles away.

Wednesday, 25 February 1987

I had slept poorly. The loss of research time was a constant frustration, and the pressure of the upcoming lecture only added to my tension. It was a relief to see the dawn come bright and sunny, a threatened snowstorm deferred for at least another day. With any luck I'd be home by midnight and back to lab work first thing Thursday.

I met Shriver for breakfast and we went over the day's schedule. It had already been decided that, in addition to giving a lecture, I was to continue my series of meetings with various faculty members in chemistry and geology. Like a zealous impresario who had successfully accomplished his mission, Shriver informed me that I was booked almost solid. In fact, there was only one late-morning slot unfilled.

"Is there anyone else you'd like to meet?" he asked.

I recalled that someone in theoretical physics at Northwestern had worked on predictions of hydrogen behavior at high pressure. That was, after all, my big Lab project—at least before the superconductor came along. So I asked who it might be.

"Could you be thinking of Art Freeman?" he suggested.

It all came back to me. Freeman's novel ideas about molecular metals were consistent with high-pressure spectroscopic data obtained at the Lab. "Yes, he's the one," I said. "Could I possibly see him today?"

Shriver made the necessary appointment, though it seemed to take some twisting of arms, and I proceeded to do my rounds.

An intense and exhausting sequence of conversations with members of the chemistry faculty followed. Each scientist was deeply involved in his or her own small corner of the scientific adventure, and I was treated to one fascinating glimpse after another—the surface chemistry of ball-bearing lubricants, the electrical properties of polymers, the crystal structures of porphyrin molecules—all significant problems but all so different from my own. I was frustrated time and again, however, as each chemist asked about my latest research. I'd never before kept secrets about my work. At the Geophysical Lab we always share data with anyone, even months before publication. That openness has always been a basic tenet of academic scientific research. So while I desperately wanted to talk about our superconductor problem, that was one subject that was strictly off limits. Instead I described past high-pressure projects that seemed to me trivial and boring compared to The Experiment.

Then it was time to meet Art Freeman. Prior to our meeting I got the mistaken impression that he was a bit reclusive. Judging from Shriver's phone conversation Freeman was reluctant to see me, and even though the Northwestern chemistry and physics departments share the same large laboratory complex, Shriver seemed unsure where to find his office. With the aid of a building directory and floor plan we finally found the right room.

Arthur Freeman was not what I expected. Theoretical physicists, whose work is accomplished with paper and pencil, a computer terminal, and the mind, often have a distinctive look about them. Rumpled clothes and disheveled hair are the general rule. But Art Freeman was neatly attired in a conservative jacket and tie, his thinning hair and graying beard were well trimmed. He is not a tall man, but his muscular body and tightly controlled movements give him an air of authority. A sober and unchanging expression revealed little of his thoughts as I was introduced.

I was confident that we shared common interests, but in his office Freeman was remote and uninterested and his annoyance at being bothered was easy to sense. I congratulated myself for successfully concealing my own frustration at being away from the x-ray lab. And, after all, what could Freeman be thinking about that matched the importance of my interrupted superconductor research? So I began to describe my high-pressure hydrogen experiments and asked his opinion of an anomaly in the experimental results.

"We're not working on that problem anymore." It was an abrupt response, almost a dismissal. He added, a little more courteously, that he'd be happy to give me copies of his papers and left the room to gather the reprints.

My eyes wandered about the small office with its stacks of reprints and preprints, shelves of esoteric physics texts, and an equation-strewn chalkboard. I absentmindedly read the seemingly random scrawls on the board—and I was astonished at what I saw:

$$Y_{1.2}Ba_{0.8}CuO_4$$

How could it be? In plain view for anyone to copy!

"Where did you get that formula?" I demanded of Freeman when he returned.

He blanched and dashed to the board, erasing the cryptic symbols with violent swipes. "No one's supposed to see that!" he said with great concern evident in his voice. "I was just discussing it with my student."

The board was blank, but he couldn't erase the symbols from my mind. To the uninitiated it was just another arcane chemical formula, but to me it clearly spelled superconductor.

"But it's Chu's secret superconductor," I insisted. "I'm working on the crystal structure. How did you get it?"

"Shhhhh!" whispered Art, who was by then peering out the doorway, his eyes apparently searching the outer office for eavesdroppers. Satisfied that we were alone and unheard, he closed the door, turned, and looked me in the eye. His expression was strangely intense, his stare penetrating. "I'm working with Paul, too. We've been doing energy band calculations. I *must* have that structure! Now, tell me everything you know."

Knowledge is power in this business and I knew something that Freeman needed in his own work. Nothing in science is valued, honored, respected—or feared—more than another's knowledge. So, with hydrogen forgotten, we set to talking about yttrium, barium, and copper.

But I was faced with a dilemma. All morning I had been avoiding sharing this information with other scientists. Chu had admonished us to tell nobody about our progress so how could I justify informing Freeman? But he was so insistent that I decided to level with him.

"Paul made us promise not to release the information. He knows our latest results. It would be better for you to go through him." What else could I do?

After some debate on the point Art agreed to call Paul in Houston. His success at getting right through, past Paul's protective battery of secretaries and lab assistants, gave considerable credence to Art's claim of partnership. After a few words he handed me the phone.

"Hi, Paul! I'm here in Northwestern to give a talk. I just met Art Freeman who claims he's working with you. I don't remember your saying anything about that. What's the story?" I'm afraid I sounded a bit abrupt, feeling that my membership in Chu's exclusive superconducting club had become a lot less exclusive if such a chance meeting led straight back to the Houston lab. Did everyone know about it?

Chu, perhaps sensing my concerns, was quick to explain. "It's fine, Bob. It's OK. Art has been working with us for the past two months on theory."

That wasn't good enough. I needed explicit approval before I would divulge any information.

"But should I give him our composition and structure results? He says he needs them for his calculations."

Once again Paul reassured me. "Yes, it's all right. You can tell him anything you know about the sample. Do you have anything new?"

I assured Paul that our tentative perovskite model for the black compound was still valid, but that no new data on the green phase had been obtained since we talked on Monday. I then handed the phone back to Art who ended the conversation quickly. He joined me at a table and we began to review what the Geophysical Lab group had deduced in the past few hectic days. I described our laborious separation of the black and green compounds, which I emphasized were intergrown on a very fine scale. The idea of two intergrown phases did not surprise the theorist, who knew of such behavior in other synthetic mixes. Freeman also seemed particularly pleased about the possibility that the black component had a perovskite-type structure. Such a simple atomic arrangement was quite amenable to theoretical treatment. We talked at some length about possible black-phase compositions until the rather limited extent of my knowledge was clear to him.

"Now tell me about the green phase."

"We really haven't done anything with it," I explained. "The crystals are so small."

"But you *have* to solve that structure! It could be the key! Paul thinks the superconductivity may be an interface phenomenon." Freeman began to explain Chu's idea that superconductivity in the new group of high-temperature oxides might result not from the black or green compound alone, but rather from the special atomic structures that occur only at the grain boundaries between two phases. If so, then we had to know structural details of both the green and the black phases.

"How soon can you fly back to Washington?" he demanded.

Actually, I was heading back pretty soon. My lecture was scheduled for 4:00 P.M. and I was planning to catch a plane back at 8:00 P.M. That wasn't soon enough for Freeman, however.

"Can't you cancel the lecture? You shouldn't be here! Your lab work is much too important." Freeman was insistent.

I knew how he felt and in retrospect he was probably right. And I, too, wanted that structure solved. But I also felt compelled to honor the Northwestern commitment. I was still defending my decision when Duward Shriver came to the rescue, ready to usher me to a lunchtime appointment. Relieved, fascinated, and as yet unable to interpret the nuances of our conversation, I gave the theorist a parting handshake and assured him: "We'll be in touch." I was sure that was not the last I'd hear from Art Freeman.

Hazen and Freeman—experimentalist and theorist—so different and yet so interdependent. For the time being I held the upper hand; there was nothing Art could do to help me solve the structure riddle. In fact, Art could do nothing at all on the new superconductor until I gave him that structure. His calculations were completely dependent on knowing the composition and atomic arrangement. But once those details were known Freeman would have a field day, for it is the theorists who tell us the why of superconductivity. Why do the electrons flow freely? Why is the critical temperature so high? How can T_c be raised even higher? It is the theorists who point the experimentalists in exciting new directions. So it was with good reason that Paul Chu had added Art Freeman to his extended research team.

With that curious encounter I concluded the morning meetings. Lunch eaten, the afternoon was spent in the more familiar world of the geology department, located in an old, weathered stone building at the opposite end of the lakefront campus. I was introduced to more new faces, more glimpses of research, more scientists from whom the stupendous superconductor discovery must remain a secret. I tried to pay attention to my hosts, but my mind kept wandering back to the extraordinary conversation with Art Freeman. The afternoon did not pass quickly.

Lecture time finally approached. I was given a few minutes to collect my thoughts. I hadn't really planned what I was going to say, even though I'd known about the lecture for months. But the topic was one of my favorites—the behavior of crystal structures at high pressure. Like most earth science speakers I just selected a sequence of sixty or seventy slides and talked about them as you would about old friends. With the exception of my propensity for pushing the wrong button, thereby going backward instead of forward through the slides, things went reasonably well.

I was happy to see Art Freeman in the audience as I began to speak, and was even more pleased when he came up afterward to congratulate me. He was relaxed and friendly, and he wished me good luck in my superconductor structure studies. I promised to keep him informed of any developments and we parted with a warm handshake.

It was a special pleasure to see my old friend, University of Chicago mineralogist Paul Moore, sitting near the front of the audience. His hair was grayer than I'd remembered, and he had put on a bit of weight, but his active mind was in high gear as always. I was sorely tempted to tell him the details of our superconductor work. As a brilliant crystallographer and the acknowledged expert on solving difficult new mineral structures, he would have been a perfect person to probe for ideas about the probable perovskite variants in the yttrium-barium-copper system. But once again my hands were tied. I couldn't even share the superconductor secret with my friends.

* * *

It was with some relief that I finally concluded my Northwestern visit and left for the airport. Everything looked good for my return flight to National Airport in D.C. The weather was clear at both ends and the United aircraft was waiting on the tarmac, fueled and ready to go.

But nothing should be taken for granted at the giant Chicago airport, where delays are apparently a way of life. My 8:00 P.M. plane was scheduled to arrive in Washington by 10:30, in plenty of time to avoid the 11:30 P.M. antinoise curfew on incoming planes. But airport congestion delayed the flight first to 8:15, then to 8:35. My agitation was not eased by the seasoned business traveler in the adjacent seat, who began to recite a litany of O'Hare horror stories.

At last, forty-five minutes late, we boarded the plane, taking off just in time to reach Washington within the legal limit. I was exhausted, but home was not my first destination. Art Freeman's exhortations about the green phase had done their damage; I knew I wouldn't be able to sleep until I had mounted the largest available crystal and begun the x-ray process anew. If I followed the right route, the Lab was on the way home from the airport anyway.

It was just after midnight when I phoned home to assure Margee that I was safe and at work. She was not exactly happy at the decision, but she wasn't surprised either. By this time my nocturnal habits were all too familiar. After a few sympathetic questions about the lecture and flight home she left me alone to the quiet confines of the x-ray lab. The diffractometer with the black crystal was grinding away, but I did no more than glance at the output to make sure things were in order. For the time being the perovskite-like black compound was forgotten. I was after the green phase this time.

The crushed sample was still preserved in oil on the glass slide, but after a quarter hour of searching it was obvious that no large green crystals were to be found. I selected another, somewhat larger chip from Paul Chu's original vial of superconductor. This sample I also crushed in oil between two glass slides, though with less pressure than before. Larger fragments resulted, but all of them seemed to be composed of many grains—and we had already learned the headaches of multiple grains! It took nearly an hour of peering down that microscope, my eyes frequently blurring from the intense staring into the lens, but I finally located a tiny, 30-micron irregular green crystal. The shape was horrendous for crystallographic work. It was a deformed triangular flat plate that actually appeared to have a hole in its middle. But I did not think I'd find a better specimen that night.

With the greatest of care I mounted the crystal; to lose it at this point would have been devastating. And then the big question: would I be able to find any reflections? I turned on the x-ray generator and started the first Polaroid x-ray photograph, which would take at least a quarter of an hour to expose properly.

It was now 3:00 A.M. and I had time to kill, so I called the Brookhaven synchrotron. Sure enough, Larry Finger and Dave Mao were hard at work, squeezing every possible minute out of the dwindling high-energy beam time. They were still struggling to align a high-pressure cell in the narrow beam of synchrotron x-rays. We shared stories of our recent progress and plans for the next couple of days, and I think we all felt a kind of camaraderie at that hour, when most people are not fiddling with tiny crystals and beams of x-rays.

The first Polaroid film was not an overwhelming success, but a faint pair of spots was proof that single-crystal diffraction was possible on the green phase. There was little point in doing more x-ray work until the glue holding the green crystal in place had dried, so with a feeling of relief and exhaustion I decided to shut things down and head for home.

There is a dead hour in Washington. I know, because I was driving home at 3:50 in the morning and there was not a soul about. No cars, no police, nothing. The night was bitterly cold and clear; I kept the window partway open to keep from dozing off. Images of the green-phase photograph, with the tantalizing weak spots, kept recurring in my mind.

I was almost to the D.C.-Maryland line, about three miles north of the Lab on Connecticut Avenue, when I saw her. A desperate brown-haired woman of perhaps twenty-five or thirty was frantically waving down my car. Something was clearly not right. She stood in the road coatless, in a thin cotton blouse, and she looked half-frozen. Thoughts of the superconductor were temporarily suppressed. There was no choice, I felt, but to stop and help with the apparent emergency.

Without any formalities she got into the front seat. "I have to get to the Royal Warrant! Please hurry! I can't find the keys."

Having expected some kind of medical crisis, I was puzzled by her reference to a restaurant. Though obviously chilled to the bone, she seemed to be more concerned about a lost set of keys.

"Please take me to the Royal Warrant," she repeated. "I'm the night manager and cook. I've lost my keys and there are customers still there. I don't know what to do."

I didn't especially feel like turning around and driving the three miles back to the small restaurant which, ironically, was located just a block from the Lab. On the other hand, I couldn't very well throw her out into the night, so back downtown we went. I was being drawn back toward the x-ray lab; would I ever escape the green phase?

The trip took only a few minutes. We pulled up to the front door of the restaurant and there in the lock, in plain view, were the missing keys. At least the strange adventure had a happy ending. I accepted the woman's offer of a beer which she poured at the nearly deserted bar. She thanked me, rather casually I thought, and asked what I did for a living.

For the first time that day I decided to respond with what was *really* on my mind. I'd been repressing my thoughts for hours and it was tremendously satisfying to blurt it out.

"I'm a research scientist and I'm studying a new superconductor."

She had been playing with her keys, but stopped suddenly as she repeated the word. "Superconductor. So what's that?"

I proceeded to tell her about Paul Chu's discovery and our work. I don't think she was impressed because after a couple of minutes she excused herself and quickly disappeared into the kitchen.

That's when the doubts began. Had I said too much? Was this a setup? She couldn't be a spy, could she? The long hours of work and the dry night air had made me thirstier and perhaps more paranoid than I'd realized. I finished my drink quickly and left without saying good-bye.

Paul Chu had entrusted me with the superconductor secret, and in my fatigue I had betrayed his trust. As I once more headed north on Connecticut Avenue I resolved not to make the same mistake again.

Chapter 5.

The Green Phase

Thursday, 26 February 1987

Margee was much more upset by my social indiscretion than with any scientific faux pas. As she saw it I had picked up a girl, gone to a bar with her, and gotten home in the middle of the night. She'll put up with overtime in the interests of research but not for extracurricular activities.

It was too late to have much more than a halfhearted confrontation, but as I got into bed I assured her that I loved her and was glad to be home. Her goodnight was much more emotional—something to the effect that I was impossible to live with and if I so much as put one toe across the invisible line down the middle of the bed I'd be one sorry scientist. When I woke up at 9:00 A.M. she was gone.

We'd had confrontations before. A few memorable fights had even lasted through the night and well into the next day before one of us had made a friendly overture or inadvertently laughed. But this occasion was disturbingly different. Margee had slipped away noiselessly, without a word or even the shadow of a touch. And now the house was strangely silent. Where had she gone?

I roused myself, determined to find her, but even before I had gotten out of bed the front door slammed shut with such force that the paintings on the bedroom wall trembled. In the brief time it took me to realize that Margee had actually left and was now back, she appeared at the bedroom door. She was still angry but she seemed to have switched targets. The school bus drivers had gone on strike and for the last hour and a half, while I had been sleeping, she had been battling rush-hour traffic.

I responded sympathetically and Margee soon calmed down. With the kids safely deposited at school, she suggested we relax and share a big breakfast. "I want to hear all about your trip," she added.

"I don't have that much time." I must have sounded remote because her tone of voice changed suddenly.

"What do you mean? You've been working for two days straight."

"I know, but I have to get to the Lab. I picked up some useful information in Chicago and have to get to work on it." By this time I was in the shower and had to talk loudly to make myself heard over the water. "I'll just grab a quick bowl of cereal on my way out."

There was no response for a minute or so. Then she drew back the shower curtain and stared at me. Her voice was hard. "OK, I understand. But I hope you have your goals in life in the proper balance." And then she turned abruptly and walked away.

I didn't know what to say. I wanted to be with her, but for the moment nothing was more important than this experiment.

The x-ray lab seemed deserted when I arrived shortly before 11:00 that morning. Larry Finger, Rus Hemley, Charlie Prewitt, and Dave Mao were all at Brookhaven involved in synchrotron research. I continued to miss being with them, sharing the camaraderie of the group effort, but I knew I could contribute more in Washington, working on the green phase.

Though there were several phone messages sitting on my desk and a four-inch stack of letters, journals, and manu-scripts awaiting processing in my mailbox, I went straight to the business of x-raying the superconductor. I knew from the first green-phase photograph that we could detect useful diffraction effects, but it was essential to have at least one more photograph with the crystal rotated 90 degrees from the first. Three-dimensional atomic structures cannot be solved without three-dimensional diffraction data.

The process took longer than I'd hoped. For several months our Polaroid x-ray camera had periodically eaten film.

Now, suddenly, the camera seemed to be getting hungrier. Each x-ray picture was taken with a separately loaded, postcard-sized packet of type-57 film. A complex loading mechanism allowed us to remove a light-tight envelope from around each film packet, expose the picture, replace the envelope, and develop the picture. All of our film failures occurred, naturally, during the last step—after the sometimes lengthy exposures had been taken. The result was wasted x-rays, useless films, lost time—and frustrated researchers.

True to form, the camera worked fine until I tried to develop the shot. The patented developing chemicals should have automatically spread evenly over the film surface. Instead they wound up in messy clots and smears inside the works. The x-ray photo was a total blank and I had to take the cassette completely apart, thoroughly clean off the brownish caustic goo, and start over. My second attempt at an x-ray picture was not perfect—the bottom third of the image was blank this time—but the photograph would do. Like the picture of the previous day, it revealed two faint spots. I had four spots in all—just enough to go to work on the diffractometer.

It was at that point that my work on the green phase began to pose a problem unrelated to solving the unknown structure. I needed a second diffractometer and that meant bumping one of our two crystallographic postdocs, Nancy Ross or Ross Angel. Both of "the Rosses," as we referred to the young engaged couple, were bright, energetic, and ambitious. Both had come with Charlie Prewitt from the State University of New York at Stony Brook, where they had become his protégés. And both relied on precious diffractometer time to collect data and establish reputations at this critical career stage.

After only a few moments of deliberation I settled on trying to enlist Nancy's aid. First, she had enjoyed more x-ray time than Ross since their arrival; she already had data for one paper and a second was well under way. Second, she was an expert on perovskites and I had already resolved to seek her help on the black phase. That problem would compensate in part for her lost data collection time. And third, her sunny disposition and even temper, in contrast to the intense personality of Ross, made it unlikely that she would protest my move or make me feel more guilty than necessary.

I found her in the pressure calibration laboratory, adjusting the laser intensity prior to checking the pressure in her diamond cell. She had been investigating the high-pressure crystal structure of the mineral stishovite, a dense form of silicon dioxide, the same composition as common quartz beach sand. As soon as she saw me she made a motion to get up.

"Hi, Bob! Do you want to use the system? I can wait." Her cheerful greeting and spontaneous offer didn't make me feel any better about my mission.

I motioned her to resume her work and proceeded to tell her the whole Paul Chu story, describing our preliminary structure and composition research and emphasizing our current progress on the perovskite-like black phase. Then I told her the bad news.

Nancy immediately recognized the importance of the superconductor opportunity, and she had already figured out that two phases meant two diffractometers. Her positive response was quick and enthusiastic—and appreciated. I promised to fill her in on the detailed results as soon as she had finished her task.

I bounded back up the stairs to the x-ray lab and immediately transferred the crystal from the x-ray camera to the diffractometer she had been using. It took only a minute to center the tiny green speck using the cross hairs of an alignment microscope. I shut the doors of the protective x-ray enclosure, turned on the x-rays, and drove the four diffractometer arcs to the expected angles for the first x-ray maximum. I found the first reflection easily. That spot centered, I went to the second. In spite of their relatively low intensities—only about 100 x-ray counts per second—I found the four peaks with little trouble. It took less than an hour to center them all.

The next step, just as it had been with the black phase, was to implement the auto-indexing routine to find the unit cell. I typed in the data for the four reflections and the computer responded almost immediately:

SINGULAR MATRIX!!!

I had made a stupid mistake. The four reflections that I had centered all lay close to one plane of the crystal. My x-ray data thus came from only two dimensions—a plane—

whereas the unit cell is three-dimensional—a box. Without additional x-ray information on reflections out of that original plane it would be impossible for the computer to calculate the size and shape of a three-dimensional unit cell. I would have to find more spots, and that meant more Polaroid photographs. I had attempted to do things "quick and dirty," and I had failed.

I resolved to obtain a complete suite of eight Polaroid shots taken by rotating the crystal in 22.5-degree increments. In that way virtually all of the strong diffraction effects could be seen. With two pictures in hand there were only six to go.

I transferred the crystal back from diffractometer to camera. But the Polaroid camera was voracious and I did not have a fun afternoon. One picture would develop perfectly, but the next film would jam or the developer packet would break or half the film would be unexposed. I spent four hours—more than twice the normal time—to take fourteen exposures, only six of which were usable. It was a maddening waste of time, but there was absolutely nothing I could do about it.

During the fifteen- to twenty-minute lulls while each picture was exposed I introduced Nancy to the black-phase problem. We knew that the black-phase structure was based on a cubic perovskite with a 3.9-angstrom unit-cell edge length. In the simplest perovskites that tiny cube repeats over and over again. But most perovskites are more complex, with clusters of two or more different cubes forming the repeating unit cell. In these complex perovskites the 3.9-angstrom cube is called the subcell, while the repeat group of several cubes is called the superstructure.

A 3.9-angstrom cubic perovskite has strong x-ray reflections that are rather widely spaced. Perovskite superstructures are revealed by the presence of additional, weaker reflections located precisely between the strong subcell reflections. If the superstructure consists of two subcells along one direction, for example, then there are twice as many x-ray peaks in that direction, usually alternating strong-weak-strong-weak in intensity. A tripled superstructure yields three times as many reflections, equally spaced and with a strong-weak-weak-strong-weak-weak intensity pattern. We had to search for these extra telltale reflections.

Throughout the previous forty-eight hours the diffracto-meter with the black crystal had searched systematically for every possible x-ray spot that might be present in a cubic unit cell with 7.8-angstrom edges—a perovskite variant in which all three 3.9-angstrom axes were doubled. Many perovskites have such double-cube superstructures and searching for the 7.8-angstrom cell was thus a logical next step in our attempts to solve the black-phase riddle.

Though less than half done with its automated task, the diffractometer had sampled and recorded more than 400 different potential reflections—more than enough data to tell us if the black phase had a doubled perovskite unit cell. A numerical record of each reflection is stored on magnetic tape and a computer converts those data into a visual format on a graphics computer terminal. We examine each reflection for proper peak shape and the total intensity of x-ray diffraction. X-ray intensities are related to the position of atoms in the unit cell, and thus they provide the key to solving an unknown crystal structure. In the case of the black phase Nancy and I were especially eager to see if any of the observed reflections indicated a doubling of the simple 3.9-angstrom perovskite cube.

We walked down the corridor to the room where the tape's data were read into the Lab's computer. We decided to use the graphics terminal in Larry's office because of its proximity to the x-ray lab. I typed a few commands and within seconds the first peak was on the screen. It was ugly. I had become spoiled working on gem-quality minerals like garnets, olivines, and spinels. Those crystals were near perfect, with sharp, clean diffraction and whopping big intensities. A good x-ray peak has a silhouette that soars like a futuristic sky-scraper; our peaks looked more like melting igloos. The maxima were broad and weak with uneven backgrounds. But what could we do? Throughout the early afternoon Nancy and I plowed through 400 reflections, groaning periodically at an especially nasty-looking peak shape. It was obvious that the black crystal was far from perfect and that it would be difficult to solve the structure based on these data. Nancy resolved to find a better crystal.

As we examined the reflections we looked carefully for a doubled perovskite unit cell. Almost all of the strong reflections—those corresponding to a 3.9-angstrom-cube subcell—were present. The peaks were poorly shaped but they were there. But as for signs of a doubled supercell, *nada;* there were no extra reflections. If the perovskite had a superstructure, it wasn't a multiple of two subcells. Of course, that still left tripled or quintrupled subcells, which are also common in the world of perovskites. We would have to check out those possibilities.

As Nancy commenced her search for a better perovskite crystal I returned to the green crystal, which was by now a seasoned traveler. I transferred the sample back to the diffractometer and used the new suite of photographs to center quickly several more weak reflections, making sure to locate anything out of the plane of the first four spots. It was again time for auto-indexing:

NO SOLUTION

Didn't this stupid program ever work? I abandoned the auto-indexing routine and resolved to use Larry's new ANGLE program that had helped so much on the black phase. It took only a couple of minutes to type in thirteen sets of four angles and print out the results.

It was déjà vu. Telltale low angles—several less than 10 degrees—were a sign of another multiple crystal. In fifteen years of crystallographic experiments I had never worked on a multiple crystal, and now there were two in one week. Paul Chu's sample had evidently formed quickly, yielding many small crystals that grew locked together. What we had hoped were single crystals of the green and black compounds had turned out to be several crystals stuck together. That made our x-ray analysis much more difficult because we had to decipher which x-ray spots came from which crystal.

But in spite of the multiple crystal problem, the ANGLE output did reveal something interesting. The thirteen reflections that I had centered appeared to originate from four different crystals. There were four distinct sets of reflections; each set was distinguished by angles of 76, 84, and 85 degrees between the three x-ray peaks, and in each set the atomic plane spacings were 2.9 and 3.0 angstroms. There was

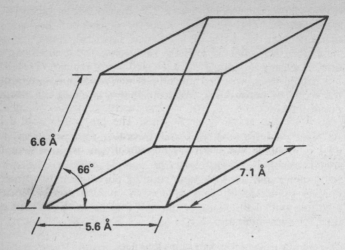

FIGURE 3. *At first I thought the green-phase crystals were mono-clinic, a symmetry that implies a unit cell with one angle different from 90 degrees. The green phase appeared to have a 66-degree angle between two axes.*

nothing familiar or special about these specific distances and angles, but the repetition of the pattern indicated that each set of three x-ray peaks came from a similar crystal. A quick analysis showed that the unit cell was "monoclinic," a common crystal symmetry in which the shape of the unit cell is like an old tippy building, with a perfect rectangular base but angled sides—a rectangular leaning Tower of Pisa. With that information I could begin a data collection.

There were other aspects of the unit cell that I probably should have checked—doubled axes or higher symmetry or other variations—but it was already past 7:00 in the evening and after the previous exhausting night I was content to start a quick data collection based on the monoclinic unit cell and call it quits. There would be time enough to improve the cell the next day.

Friday, 27 February 1987

The fast green-phase data collection, begun the previous evening, was complete when I arrived back at the Lab at 8:00 A.M. A quick scan of the computer output revealed more than a dozen reasonably strong diffraction spots. I chose a representative set of eight peaks and instructed the computer to measure the unit-cell dimensions. The program would drive the diffractometer arcs, precisely center each reflection, and then compute the answer automatically. I could ignore the green phase for several hours.

Nancy had succeeded in finding three or more possible black-phase crystals. Each had to be mounted, photographed, and oriented in order to determine which one—if any—was better than the first. I was happy to leave that problem with her for the time being.

In her efforts to understand the black phase, Nancy decided to consult with her fiancé and sometime collaborator Ross Angel. Ross is an experimental mineralogist with special interests in the esoterica of crystal structure analysis. He came to the United States from England on a NATO postdoctoral fellowship to work with Charlie Prewitt at Stony Brook, and then migrated south to the Lab when Charlie did.

Nancy thought Ross would be surprised at the remarkable superconductor news. But he surprised her.

"Actually, I've known about it for a week," he told her. "I was in the x-ray lab last Friday and overheard the commotion. But Larry made me promise not to tell anyone." Ross was evidently a lot better at keeping secrets than I.

Like the rest of us, he had instantly recognized the superconductor project as compelling and important work. Ross also understood the proprietary nature of the research so he was hesitant to intrude. But now with Nancy's encouragement, he came into the x-ray lab just as I was about to leave and offered his help.

"I heard the news," he said. His words were succinct but were softened slightly by the lilt of the Queen's English. Ross was dressed as usual in a coat and tie, a strangely conservative outfit for a twenty-eight-year-old and one that contrasted jarringly with his long, dark, sixties-era hair. Ross occasionally adopted an intense, argumentative style of communication that struck me as a cross between the language of a Cambridge debater and the bluster of a Manchester rugby player. But that day he was easygoing and very friendly. "I'd like to help," he said simply.

I was more than happy to have him working with us. Ross was particularly adept at the solution of crystal structures by direct methods, a computational approach that might work well on solving the atomic arrangement of the green phase, so I showed him the status of my unit-cell analysis. Soon, I promised, he would have a set of data. But he was so eager to get started that he decided to get a jump on things by analyzing the hasty set of x-ray intensities that I had obtained the previous evening.

I was delighted at the enthusiasm of the two young Rosses, and reckoned that my job was becoming a lot easier. Over lunch I actually had time to take care of some backlogged paperwork that had been turning my desk into a disaster area.

I like a neat desk. I'm not as fastidious as some of my crystallographic acquaintances who leave paperwork in perfect piles, symmetrically disposed about their blotters and precisely aligned parallel to the edges of their desks, every scrap in place. But I do like a neat desk.

After a week of superconductor research the top of my beautiful old oak desk was almost lost from view. Computer output, diffractometer logs, and an assortment of other superconductor detritus were interspersed with unopened mail, unread journals, and unanswered phone messages. I tackled the organizational chore with gusto, and managed to control the flood at least for another day or two.

After lunch I went to check on the green-phase unit-cell measurements. I watched the results being printed out automatically as the second hand swept by 1:00 P.M. The computer reported a unit cell with its three edge lengths of 5.7, 6.7, and 7.1 angstroms and angles between these edges of 90, 90, and 65 degrees, respectively. (Two 90-degree angles are diagnostic of the type of symmetry called monoclinic.) I noticed with satisfaction that our experimental errors on the unit-cell measurement were quite small—less than one part in a thousand.

That meant that we knew the unit cell's size and shape extremely well.

I was almost set to begin a careful data collection but I had a nagging suspicion that I'd made a mistake. In my haste to assume the monoclinic unit cell I never checked for obvious alternative symmetries. Once again I invoked the auto-indexing program. I typed in data for some of the reflections that the diffractometer had found overnight. I expected the computer to simply confirm the monoclinic unit cell that I had been using.

Auto-indexing worked for what seemed like the first time in ages. The new, correct cell appeared on the console:

7.11   12.16   5.65   90   90   90

Three 90-degree angles! The green phase was not monoclinic with two 90-degree angles; it was orthorhombic with three right angles. Orthorhombic unit cells are rectangular prisms; the green-phase orthorhombic unit cell was shaped something like a shoe box. My mistake was an easy one to make, and fortunately it was easily rectified, because the monoclinic and orthorhombic cells were closely related in size and shape.

I called Ross into the lab and showed him the new results. Having gotten nowhere with the erroneous monoclinic cell, he was happy about the higher orthorhombic symmetry because it would make solving the structure a lot easier. Before we did any more x-ray work, however, we had to find out if we had a previously described structure. Virtually every known crystal structure is listed systematically in *Crystal Structure Determinative Tables*, an indispensable reference work in the x-ray lab. All orthorhombic structures are listed according to axial ratios. We had measured axial lengths of 5.65, 7.11, and 12.16 angstroms; the ratios of these numbers were a kind of fingerprint for the atomic arrangement of this compound. Nothing fit. The green-phase structure evidently was unrecorded; it was ours to solve. By 2:30 P.M. we had begun a long, careful collection of intensity data, the key to deducing the atomic arrangement. Soon Ross would have decent numbers to work with.

* * *

In the meantime we headed upstairs to tell Charlie Prewitt about our progress. He was having a hard time finding any logical fit to the Houston x-ray powder pattern. Our new information on both the black- and green-phase unit cells would make his task much simpler. Now Charlie could calculate a complete set of expected powder lines from our cell data, and then compare those lines with the ones recorded on the Houston chart. In particular we wanted to know if there were any strong peaks that could not be explained by the known black- or green-phase unit cells. That way Prewitt could tell us if there was a third, as yet undetected compound in the superconducting sample.

He was puzzling over the pattern when Ross and I arrived at his office. "There's something really strange about these patterns," he told us. "If I fit the low-angle peaks, the high-angle peaks seem to be off, and vice versa. I'm beginning to suspect a mechanical error in the Houston diffractometer, but I can't believe the Houston machine is that far off."

Charlie had been working with the x-ray data collected by Ken Forster at Houston. At the time we had no idea that the Houston x-ray pattern had been deliberately altered and that we had inadvertently been sent the doctored data. Without knowing the correct values for the green and black unit cells there was no way for him to decipher the powder pattern. But, with the new unit-cell information in hand, it only took Charlie a few minutes to discover the systematic angle error, correct the pattern, and identify almost every peak. Without a doubt the green and black compounds together accounted for virtually all of Paul Chu's sample. There were only two phases.

Dave Mao was also back to work on the superconductor problem. He had returned Thursday night from a discouraging session at Brookhaven, where the Lab team had spent all their time struggling with the sensitive alignment of diamond pressure cell, x-ray beam, and detector to no avail. Not one bit of hydrogen data had been obtained and time was running out. Rus Hemley, Larry Finger, and Andy Jephcoat, one of the Lab's mineral physics postdocs, were still at it; but Dave de-

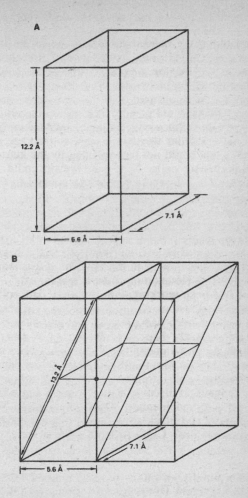

FIGURE 4. *The green phase turned out to be orthorhombic, not monoclinic as originally thought. In orthorhombic crystals every unit-cell face is a rectangle. The shape of the green-phase unit cell was similar to a shoe box (top). The simple relationship between the erroneous monoclinic cell and the correct orthorhombic cell is shown at the bottom.*

cided to come back to attack the composition problem again. This time he had help.

Chris Hadidiacos, microprobe jockey extraordinaire, made the difference. Chris, officially known as the Geophysical Laboratory's "electronics engineer," knows more about the Lab's scientific hardware than most of the scientists. If something is broken or needs a state-of-the-art modification Chris, oscilloscope in hand, is usually the troubleshooter. With a tall, athletic build and an easygoing, confident style his appearance makes us feel a lot better when critical lab apparatus malfunctions. He is like a respected family physician and we call for him often.

While Dave was in Long Island Chris had made two important improvements to the microprobe. First, he located chemical standards that would allow us to measure how much of each element was in each grain. And second, he adjusted the size of the electron beam to focus down to a smaller spot. After some consultation with his right-hand man, David George, Chris also prepared the critical sample mount. He embedded a portion of the fine-grained superconductor sample in epoxy cement on a circular glass microprobe slide. Once the adhesive was dry he cautiously polished the assembly, barely exposing portions of the green and black specks, which were indistinguishable from each other in this configuration.

This time the probe results were not ambiguous. Dave and Chris saw a range of compositions, but there were two groups of distinctly different compositions, each with a perfect integral ratio of the metal oxides. Most of the grains displayed a simple 2:1:1 ratio of yttrium to barium to copper atoms, giving a formula Y_2BaCuO_5. We assumed that composition corresponded to the green phase because most of the sample was green and because Y_2BaCuO_5 was not a perovskite-like formula.

A smaller number of grains gave an unusual 1:2:3 ratio of Y to Ba to Cu, corresponding to a formula of $YBa_2Cu_3O_{6.5}$, give or take a little oxygen. That must be the black phase. It was an odd formula for a perovskite, which ordinarily would have nine oxygens for every six metals, but, like other perovskites, it did have an equal number of large metals (yttrium plus two bariums) and small metals (three coppers).

I reviewed in my mind what we had learned so far. There

werc two phases. About two thirds of the superconductor was a green, orthorhombic compound with the formula Y_2BaCuO_5. The remaining third was a black, perovskite-related phase with the composition $YBa_2Cu_3O_{6.5}$. With unit cells and compositions in hand and x-ray intensity measurements under way, we had made significant progress. We were on the final, most difficult leg of our race; soon we would have the data needed to determine the actual arrangement of atoms.

I was eager to report our progress to Paul Chu. My call was answered by his secretary, who informed me that Paul was in California. I gave her both my work and home numbers and hoped he'd get back to me soon.

I had to leave. I had promised to pick Ben up at school by 6:00 P.M. and it was already 5:45. Nancy was in charge of the black-phase data collection on the South Huber and Ross would oversee the green phase on the North Picker diffractometer. They would call me at home if any problems arose, but I was confident that they could handle the monitoring chores.

I made it to Chevy Chase Elementary School at 6:01, just a minute late. Ben's entire fifth grade class was returning from a two-day field trip to Williamsburg in Virginia and all parents were instructed to retrieve their children at the school's parking lot. There was no sign of a bus and many other parents were waiting in their cars or congregating near the school entrance. I had obviously arrived in time.

Having rushed to get there, I thought I'd be upset at any delay. But instead I found the enforced break from Lab work quite relaxing. During the past few weeks I'd been neglecting another love of mine, the trumpet. In fact, I had some exposed solo lines in an upcoming Handel Festival concert at the Kennedy Center and I was not yet entirely prepared. So for the next half hour or so I practiced my tiny piccolo trumpet in the car.

For almost two decades I had played symphonic trumpet professionally, as much for love as for money. Throughout much of my childhood in northern New Jersey science and music had vied for attention: town band and church choir versus astronomy and rock collecting. When the time came for college my choice was MIT, not just for its well-known science curriculum, but also because of its noted symphony orchestra and proximity to the New England Conservatory of Music.

For years I had dreaded the inevitable, irrevocable decision between a career in science or music. How could I choose? Professor Martin Buerger, the formidable, dour, and reserved teacher of my freshman crystallography course, clarified my thinking. A conscientious churchgoer, he observed me playing at a Sunday morning service at his Park Street Congregational Church in Boston. The next day he took me aside and, with a mixture of fatherly advice and stern rebuke, he advised me to abandon music entirely.

"No dedicated scientist can afford to waste his time on other pursuits," he told me. "You should devote yourself to your studies, not music." There was no ambiguity in his harsh dictum, no doubt or equivocation in his tone. In an instant my choice had been made for I knew that Professor Buerger was wrong. Science and art are complementary, not rival, disciplines. Music gave me a perspective on science as a human endeavor that many scientists would never enjoy. Science could provide an insight to the physical world that few musicians ever enjoy. I resolved to be both a scientist and a musician for as long as time and ability would allow.

As I practiced it was amusing to watch the confused faces of passersby. Most couldn't quite make out the origin of the unusual high trumpet music with its ornate turns and trills. The few who did spot me were obviously torn between polite indifference and the desire to stare. I waved to those who worked up sufficient nerve to smile and acknowledge my efforts. It never hurts for a performer to reach a rapport with the audience.

After the practice session I reclined in my bucket seat, turned on the car radio, closing my eyes and listening to some relaxing chamber music. I just wasn't in the mood to mingle with the growing knots of increasingly disgruntled parents. Ben's bus finally pulled up at 6:58. Naturally my son rode in the last seat and was the last one off the bus, which seemed to disgorge its youthful freight at a glacial pace. But I had practiced and the Lab was running smoothly without me and I was looking forward to hearing my son's enthusiastic commentary on his Williamsburg adventure. I was only a little nonplussed at his perception of the educational high point—an all-you-can-eat breakfast bar that the group had savaged that morning. So much for history lessons!

Paul Chu's phone call came shortly after 9:00 in the evening. We skipped the small talk and I got right to the point. I proceeded to tell him the new and crucial information: composition, unit-cell dimensions, and the fact that we had only two principal phases. Armed with those pieces of the superconductor puzzle Paul would direct his Houston team to synthesize new samples of nearly pure black-phase and green-phase compositions. With any luck we would soon have much larger crystals to work with—and we would know for sure which phase was the superconductor.

"That's great, Bob! We'll go right to work on trying these compositions. By the way, I'm going to be back in Washington soon. I'll try to get to the Lab later this week." His tone was ebullient.

But in a pattern that was to become all too familiar, his tone suddenly darkened and he issued a warning. "You know, Bob, our article that announces the yttrium composition just came out. I've heard that people at several places are already making the superconductor; they're working on it round the clock.

"There was also a report a couple of days ago in Chinese. It was in the *People's Daily*. They also mention a Y-Ba-Cu oxide superconductor. Everyone knows the composition."

He didn't need to elaborate. Our one advantage in the superconductor race—exclusive access to the sample—was gone. Anyone with a furnace and the right chemicals could make the stuff. It was true, we did have a head start knowing the compositions and unit cells of the two phases, but it wouldn't take long for Bell and IBM and the other industrial labs to catch up. And how could we hope to compete against the giant research companies?

I promised Paul that we'd do our best and would keep him informed of our progress. I only hoped that we could live up to his expectations.

Chapter 6.

Ross Angel's Solution

Saturday, 28 February 1987

Every Washington winter has its share of warm, sunny days, a partial recompense for the relentless heat and humidity of summer. Saturday was a classic, with the thermometer passing sixty degrees by midmorning. But I was two steps removed from enjoying the balmy weather. Not only was I eager to resume work, but for a few hours I had to don my other hat—that of a musician.

Conflicts between trumpeting and research are surprisingly few. Most rehearsals and performances are held in the evenings and on weekends. The occasional weekday recording session or National Symphony rehearsal is balanced by weekends and evenings at the Lab. Only rarely, as on that Saturday morning, is pressing Lab business superseded by musical commitments.

My rehearsal was held at St. Paul's, a spacious church just one mile north of the Lab on Connecticut Avenue. A chorus amassed from several D.C.-area church choirs would perform Brahms's *German Requiem* the following afternoon. A competent professional pickup orchestra provided the accompaniment for union scale wages. The rehearsal began with a silent prayer for an ill chorister. Then we got down to work.

The first trumpet part to the *Requiem* is not difficult, with only a couple of tricky soft solo bits in the second movement and several "tacet" movements with no trumpet part at all. With so many other things on my mind I more or less played on autopilot, my music-making reflexive rather than inspired. For most orchestral situations that may in fact be the best strategy; an overinspired trumpet player can wreak havoc with orchestral balance.

Ken Lowenburg, a gifted organist and the designated conductor for the occasion, coached the orchestra, chorus, and soloists for the two-and-a-half-hour call, but in spite of strict union rules about overtime, we continued for several minutes past the usual quitting time. Grumbles were heard from several orchestra members because we knew that the close friendship between conductor and contractor might result in loss of that overtime pay. Probably sensing the growing discontent among his forces, Ken abruptly ended the rehearsal.

Restless musicians quickly moved to clean and case their instruments, but our motions were halted by the choirmaster who asked for silence.

"Please, everyone, may I have your attention," he said in an emotional, broken voice. Though eager to leave, the instrumentalists were quick to respond to his obvious distress.

"Our dear friend Charles Gilcrest has left us for a better place." It took a brief moment to realize that he was speaking of death as opposed to, say, the beach. "He labored all night to complete our concert programs, but the strain was too great. His heart gave out. Can we share in a moment of silent prayer?"

Chorus and orchestra alike were sobered, the usual post-rehearsal banter subdued. The closeness of death surely affected each of us differently, but all were reminded of life's priorities and pressures. Thursday morning's conversation with Margee haunted me. Perhaps I *had* lost my perspective. But life must go on and my immediate priorities were clear. I packed up music and trumpet and headed straight for the Lab.

The green-phase diffractometer was working smoothly, and Ross Angel was waiting patiently, when I arrived at 1:00 P.M. Our automated x-ray equipment had systematically scanned more than 200 reflections overnight; as soon as I processed the latest batch of data Ross could begin to analyze the structure. I dismounted the magnetic tape that contained the night's data and mounted a second tape "on the fly" as the machine continued to count x-rays. Ross and I examined each of the green-phase reflections to calculate their intensities (automatically) and appraise their quality (subjectively). The shapes were unusually broad and not of ideal form, but we couldn't be choosy when it came to the green phase.

It was not a complete set of x-ray data—another twenty-four hours of diffractometer time would be needed for that—but it would do for starters. Ross dashed off to his office to begin the structure analysis. He knew the size and shape of the unit cell: the repeat unit of the green phase was a rectangular box about 6 by 7 by 12 angstroms. He also knew the green-phase composition: Y_2BaCuO_5. Ross's job was to locate the positions of yttrium, barium, copper, and oxygen atoms inside the box.

Solving unknown crystal structures is as much an art as a science; it takes intuition as well as computational skill. There is no one "correct" way to solve a crystal structure. Ross realized that in the case of the green phase the key to locating atomic positions was to first find the heavy barium atoms. Barium atoms would be the easiest to pinpoint, he suspected, because they scatter x-rays more efficiently than lighter, smaller, yttrium, copper, or oxygen.

Ross began his analysis by instructing the computer to take all of the previous night's x-ray intensities and calculate "Patterson maps," diagrams that sometimes help in locating heavy atoms such as barium. The first maps had poor resolution because our green-phase intensity data were not complete. Even so, it appeared that the structure possessed a distinctive layering of atoms; the first Patterson maps indicated that atoms were concentrated in two crystallographic planes. It was tantalizing information because there are special shortcuts possible for solving layered atomic structures. But the partial data set just wasn't good enough. After hours of futile attempts to pluck out a few barium atom positions from the Patterson maps Ross gave up and went home. That night, however, sleep would not come to the eager young scientist. He spent most of the night scribbling out possible solutions to the green-phase structure.

With Ross Angel handling the green phase, I turned my focus to the black. Shortly after 3:00 P.M. I found Nancy Ross in her office, where we conferred for a few minutes. The previous evening she had begun a careful data collection of all the 3.9-angstrom cubic subcell intensities. That relatively quick experiment, which had finished overnight, employed the first black crystal. Nancy had also put aside three other promising black specks for possible study.

Our principal concern regarding the black phase was still to find a superstructure. We couldn't begin to solve the crystal structure if we didn't know the size and shape of the unit cell. So far we knew that the perovskite subcell was a 3.9-angstrom cube, and we had eliminated the possibility of a doubled cell. Nancy and I agreed that the next logical step was to determine if the perovskite had a tripled-cell superstructure. We set the diffractometer to search automatically for this large unit cell.

With both superconductor phases under the care of the Rosses I once again resolved to check on the ever-increasing pile of paperwork that continued threatening to appropriate my desk. As I returned down the hall to my office I saw Rus Hemley, just back from Brookhaven. Eager to find out the latest news, I called out and quickened my pace.

He looked terrible. A three-day growth of beard, dark circles under his eyes, and a slow, labored step were ample evidence of his exhausted state. But his face brightened when he saw me.

"Andy and I got hydrogen to 250 kbar! It was fantastic; we had to work nonstop and just finished the last data point before shutdown. We almost killed ourselves!" It was a remarkable achievement. To measure hydrogen at any pressure was a feat, but they had managed to do it at a quarter of a million atmospheres pressure. No wonder he was elated.

The irony of his last remark did not escape me as I thought of the chorister's death earlier that day. I wondered how much that last hydrogen datum was worth in human life. But Rus was young and strong and this small triumph would probably be remembered as a high point for years to come. And I had to admit that running an experiment under almost any conditions was much more exciting than the routine paperwork that occupied me for the next several hours.

I didn't leave the Lab until late. Stops at the wine store and the video club on the way home made me later still. But tonight, for once, the delays were part of the plan. Margee and I were going to spend a quiet evening together at home.

She had already fed the children and directed them upstairs when I arrived home. All that remained was to heat the spiced shrimp, dish out the coleslaw, and pour the wine. It was one of our favorite meals—a little messy, but fun to eat. Besides, it could be prolonged as long as we wanted.

Within minutes we were comfortably settled in the family room, food spread out in front of us and a video adventure movie on the TV. The content of the movie was largely irrelevant because we intended to talk through most of it anyway.

"So how was your day?" I asked as I peeled my first shrimp and savored it.

"Good," she said. "I finally finished that band article— which I hope you'll take a look at, by the way. The rest of the day I spent reading." She dipped a shrimp in barbecue sauce and ate it. "How was yours?"

"Sort of weird. Some guy in the choir at this morning's rehearsal died."

"You mean right there, at the church?"

"Yeah, just before the gig. He had been working all night on the concert program design and he just collapsed of a heart attack. It sort of struck home. Anyway, progress is definitely being made. In fact, if we keep working at this pace we might even be next week where we hoped we'd be last week. I think we almost have enough data to solve the structures." I then began a disquisition on the subtleties of green and black samples.

I must have gone on longer than I'd thought because by the time I'd finished she had pushed her plate aside and was staring at me.

"It's really a tough and all-consuming project, isn't it?"

"Yeah, I've never done anything like this before. There seems to be such time pressure. It sure isn't like good ol' geology."

"I wish I could help," she said, "but I feel so uninvolved in it. You know what I mean?"

We moved to sit closer together on the couch and enjoyed the food and the movie in silence. That's when the telephone rang.

I groaned and moved to answer it, promising that if it were

for me I'd make it quick. Famous last words. It was Art Freeman and the call must have taken at least ten minutes as I gave him data and he passed on the latest superconductor rumors. He seemed pleased with our progress, but kept asking for more black-phase data as soon as we could get it. Then he said a disturbing thing.

"Bob, I'm not sure you've heard the latest reports. Several labs have isolated the black phase and have started crystal structure work. You realize that time is getting short."

I realized that only too well. As I wound up the call I began to feel discouraged. Not only was the laboratory pressure unlikely to let up, but my one evening's break from science was already deteriorating. As I headed back to the family room, I decided not to mention Freeman's call for a while, but I should have known that Margee would have figured out who it was.

"So, that was Art, huh?" she asked as soon as I sat down.

I nodded and sighed, "The screws are turning tighter."

"Hey, cheer up, this can't go on forever."

It was a good line—I'd used it myself during the hydrogen experiments. But this time, with no obvious end to the race in sight, I wasn't so sure.

Sunday, 1 March 1987

A month earlier I had promised to play for the christening service of a musician friend's infant son. Most Sunday services last about an hour, with perhaps an extra twenty minutes for the musical prelude and postlude. But this service was a monster: eight christenings, thirty-one new church members, and a full-blown sermon in addition to the special music. Each of the baptisms and church inductions was celebrated with a separate small ceremony. For two and a half hours we sat and watched the ranks of the faithful swell.

When I finally made it to the Lab at 12:45 P.M. I was surprised to find Margee and the kids waiting. It was not a social visit.

"Bob, I need your advice. When we got back from Sunday school there was a message on the answering machine from Paul Chu. It sounded important and he wants you to call him in California. I tried to reach you but our phone just stopped working. It's dead." Margee almost never sounded this upset. Before I could reply she added in a more subdued tone: "I wanted to help and I wasn't sure what to do."

"Great—a broken phone. That's all we need! It must have gone out late this morning, then?" I asked.

"Yes, I left at about ten-thirty. Paul's message and the dead phone must have both happened after then." She looked at me expectantly.

"Well, I can't do anything until after the concert this afternoon. Why don't we call the phone company from here and see what they can tell us. Can you take care of it while I call Paul?"

She agreed to try the phone company from Larry's office while I went down to my desk to try the West Coast number. I was surprised to hear Paul answer after just one ring.

"Bob! Hi! Thanks for calling back. I've got good news. We've made some new samples that are almost a hundred percent superconducting. They are black phase." His enthusiasm was infectious.

Paul's group had made an important discovery. Following our compositional measurements the Houston researchers had mixed chemicals to the exact proportions that we had observed in the black phase: one part yttrium to two parts barium to three parts copper. The result was a pure superconducting sample; the black phase was the superconductor. Furthermore, knowing the correct composition might enable them to produce larger crystals—crystals that would improve our chances of solving the structure.

I responded with a similar tone. "That's wonderful, Paul! How soon can we get new samples?"

"We're still working on growing bigger crystals. These are very fine grained. But I'm planning to come by the Lab next Thursday. I'll bring some then."

"Terrific, I'm really looking forward to meeting you in person. We're making a lot of progress here. The green phase should be solved soon, though I suppose now that's of secondary interest?"

"It looks that way, Bob. We think the black phase is the critical one." Paul still sounded cheerful; I don't think he realized how much time we had spent working on the green

FIGURE 1. J. Georg Bednorz *(left)* and K. Alex Müller of IBM's Zurich Research Laboratory discovered high-temperature superconductivity in an oxide of lanthanum, barium, and copper. They were awarded the 1987 Nobel Prize in physics for their achievement. *(Courtesy of IBM Corporation)*

FIGURE 2. Paul Chu, September 1987.

FIGURE 3. Professor Arthur Freeman, theoretical physicist at Northwestern University, and his graduate student Jaejun Yu worked closely with Paul Chu in computing the electronic properties of the new oxide superconductors. *(Courtesy of Northwestern University)*

FIGURE 4. Professor Maw-Kuen Wu *(right)* and graduate students Chaun-Jue Torng and James Ashburn, three physicists at the University of Alabama in Huntsville, worked with Paul Chu in the synthesis of the new superconductor. The first 90 K samples were produced in the Alabama Laboratory. *(Courtesy of the University of Alabama)*

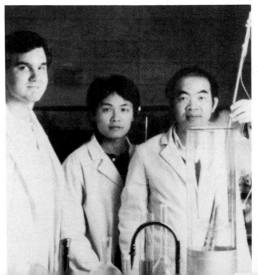

Discovery may earn billions, Nobel for UH

By CARLOS BYARS
Houston Chronicle Science Writer

Researchers at the University of Houston have grasped the Holy Grail of low-temperature physics, surpassing in a few months of intense work a goal long believed to be totally out of reach.

Their discovery of a superconductor that works at a temperature higher than liquid nitrogen could revolutionize the electrical industry, earn a Nobel Prize and be worth billions to the university.

Paul Chu, head of a team working to create new superconductors — materials that conduct electricity without resistance — said the new material operates at 98 degrees Kelvin. This is well above the critical 77-degree temperature of liquid nitrogen.

This temperature, though low by human standards at minus 367 degrees Fahrenheit, is more than high enough to allow the use of cheap, readily available liquid nitrogen as a coolant. Previously, the temperatures required for superconductivity could be achieved only by using much scarcer liquid helium, virtually precluding any possibility of widespread commercial use of superconductors.

The discovery holds the promise of major advances in the transmission and use of electrical power, including better detectors of low-level heat and radio signals, more powerful magnets and motors and new tools for medicine.

Chu's discovery was announced here and in a scientific paper accepted for publication in Physical Review Letters, considered the most important journal of physics. This was the fourth major increase in the temperature of superconductivity in the last three months, after 13 years without a breakthrough in the field.

The discovery appears to have put Chu and the University of Houston far ahead of several well-funded, highly competitive teams. At stake is more than the intense competition for scientific honors and awards. The organization that wins clear title to the basic patents stands to gain license fees amounting to billions of dollars.

Sharing credit for the discovery was M.K. Wu of the University of Alabama, who was listed as co-author with Chu of a separate paper in the same publication announcing a superconductor at 93 degrees Kelvin. Wu is one of Chu's former students.

Further advances are expected. Roy Weinstein, UH dean of natural science and mathematics, said the group has seen signs that superconductivity occurs

See SUPERCONDUCTOR on Page 5.

E. Joseph Deering / Chronicle

UH physics professor Paul Chu and associate R. L. Meng work on creating a material that becomes a superconductor at a temperature far higher than any known before.

Superconductor could garner Nobel for UH

Continued from Page 1.

above 148 degrees Kelvin (minus 312 degrees F.), and very preliminary results that that Chu's line of research may result in superconductivity at or near room temperature. Such an astounding development has not previously been considered even a possibility by researchers in the field of low-temperature physics.

Chu's group includes P.H. Hor, R.L. Meng, L. Gao, Z.J. Huang and Y.Q. Wang of the University of Houston and Wu and J.R. Ashburn of the University of Alabama at Huntsville. They are in hot competition with other researchers including groups at AT&T Bell Labs, University of Tokyo, Argonne National Laboratory, IBM, Los Alamos National Laboratory and Stanford University.

Weinstein says the competition involves three major issues. The discovery at an International Business Machines Corp. lab in Zurich, Switzerland, of the combination of chemicals used to make the superconductor; the actual production of a superconductor from that material, which was first achieved at the University of Houston; and most important, Chu's discovery of the role of pressure in superconductivity, now called the "Chu effect."

"IBM-Zurich found some promising material. They didn't know how to make it a superconductor or how to vary it to get a superconductor, but they did get everybody started in the right direction," Weinstein said. "Paul is the leader in developing new techniques for making the superconductor material."

"There is some hope that Chu will get a Nobel Prize out of this," Weinstein added. "If he does, it will be for discovering the role of pressure."

Before Chu's discovery, pressure was not thought to have any effect on superconducting materials. But Chu found

that the superconductivity temperature rises when the new materials are squeezed under high pressure.

Weinstein said this was Chu's key clue, that he was dealing with something new. But pressure is not needed to build the material, he added.

"When you apply pressure you reduce the spacing between the individual parts of the molecules. What's important is to close up the internal spaces. This can be done by filling in those spaces with smaller atoms. You don't have to use pressure. You use a different approach," Weinstein said.

One different approach is to replace one of the ingredients with another that is physically smaller. The original mixture suggested by IBM was lanthanum, barium, copper and oxygen (copper oxide). Chu now is working with strontium instead of barium or yttrium instead of lanthanum.

Weinstein says the Chu effect has revealed a whole new group of substances which are, or may be, superconductors.

Another way to reduce the spaces is to lay the superconducting material a few atoms at a time on a fine crystalline base material. As the atoms are deposited on the base, they tend to line up in the same tightly packed arrangement as the crystals of the base, thus reducing the spaces.

"If Paul had a hundred PhDs he could keep them all working on different systems," Weinstein said.

Weinstein said the university has applied for a patent on Chu's discoveries, claiming a huge variety of chemicals and processes to make superconductors.

But the secret to prodigious wealth is to turn the material into a wire. "If we can't make a wire it will be a very fine discovery but not much money," he said.

FIGURE 5. On February 16, 1987, *Houston Chronicle* announced the discovery of 90 K superconductivity. (*Copyright:* Houston Chronicle. *Reprinted with permission.*)

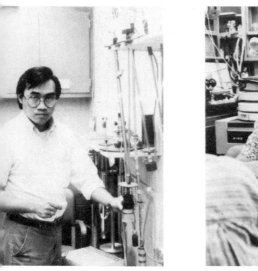

FIGURE 6. Drs. Pei-Herng Hor *(left)* and Ru-Ling Meng helped to run the University of Houston laboratory and performed many of the critical synthesis and electrical conductivity experiments of the new superconductor.

FIGURE 7. Professor Simon Moss *(right)* and graduate student Ken Forster work in the crystallography laboratory adjacent to Chu's superconductivity facility. They helped to characterize the first superconductor sample.

FIGURE 8. Geophysical Laboratory Director Charles Prewitt *(left)*, postdoctoral fellows Ross Angel and Nancy Ross *(center)*, and staff member Larry Finger *(seated)* are members of the crystallographic team who solved the structures of the black and green phase.

FIGURE 9. Drs. Ho-Kwang (David) Mao *(left)* and Rus Hemley, staff members at the Geophysical Laboratory, measured the composition and spectral properties of the new superconductor. Paul Chu sent his samples to the Lab because of his friendship with Dave Mao.

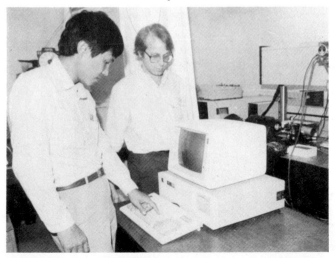

Figure 10. Ben, Bob, Margee and Elizabeth Hazen were unaware of the discoveries in high-temperature superconductivity prior to February 20, 1987, when Paul Chu sent his uncharacterized sample to the Geophysical Laboratory.

Figure 11. A cube-shaped magnet, half an inch on a side, hovers above a chilled superconducting disk. This levitation demonstrates the ability of superconductors to repel magnetic fields, a phenomenon known as the Meissner effect. This property has led to new types of electric motors and magnetically levitated trains. *(AT&T Photo)*

FIGURE 12. AT&T Bell Laboratories researchers *(from left)* Edward Reitman, Bertram Batlogg, Robert Cava, and Robert van Dover were part of the New Jersey team that learned of Paul Chu's superconductor discovery on Friday, February 27, 1987. Within a week they reproduced Chu's results, isolated the superconducting phase, and recognized the perovskite-related structure. *(AT&T photo)*

FIGURE 13. Bell Communications Research scientists, including Laura Greene and William Feldmann, were among the first to confirm Paul Chu's announcement. Bellcore researchers had synthesized yttrium-barium-copper oxides as early as January 3, but they didn't measure the samples for superconductivity until after the Houston results were publicized. Bellcore was one of four groups to isolate 1-2-3 and determine its structure prior to the March 18 American Physical Society Meeting. *(Bellcore Photo)*

FIGURE 14. Grace Lim, Robert Beyers, and Rick Savoy *(seated)* labored to characterize the yttrium-bearing superconductor at IBM's Almaden Research Laboratory in California. They successfully isolated the phase and determined its composition and crystal structure prior to the March 18 APS meeting. *(Courtesy of IBM)*

phase, neglecting the black. What was I going to tell Ross?

"I'll give you a report on everything when you come." We were just signing off when I remembered to add: "Oh, yes, our home phone is on the blink. You might not be able to get through to us there." I hoped that wouldn't cause any problems. The telephone was such a vital part of modern science. In these days of transcontinental research it could be the lifeline of an experiment.

I hurried back upstairs to check on Margee's success. Apparently things were going to be OK. Somehow our electronic burglar alarm, with its automatic phone link, had failed to disconnect the last time it reported in. The phone was working again. I walked Margee and the kids to the car, thanked her for helping me out, and waited until she pulled out of view before returning to the x-ray lab.

With assorted phone business out of the way I could at last commence what I had planned to do all along—process the next batch of green-phase data. Despite Paul's new findings, we still were committed to analyzing both phases. The diffractometer had been spewing out x-ray intensities steadily since the previous day and there were more than 400 reflections to examine. I used Larry's terminal, which processed the data more quickly than our other computer consoles, and started working on the set.

I was about halfway through the data when Ross came into the room. He looked beat and he didn't need to tell me that he had worked most of the night. I could imagine that the first, incomplete data set was insufficient to solve the structure. But in spite of his obvious eagerness to obtain the next critical reflection file he waited patiently as I examined and integrated each peak.

As I chipped away at the data I told him the news. "I just talked to Paul Chu. He's got better samples."

"Are there larger crystals this time?" Ross responded.

"Not yet. They're still trying to grow bigger ones. By the way, the superconductor is virtually all black phase." I thought Ross would be upset, knowing that the status of the green phase had just plummeted, but like any good scientist he was interested primarily in knowledge, not status. To him the unsolved green-phase crystal structure was just as chal-

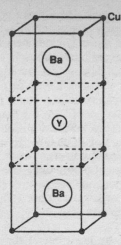

FIGURE 5. *The tripled perovskite unit cell, coupled with the 1:2:3 ratio of yttrium:barium:copper, suggested a simple perovskite-like arrangement of metal atoms. But where were the oxygens?*

lenging whether or not it had economic implications. Our conversation drifted toward the strategy for the afternoon's computing. Finally, shortly after 2:00 P.M., Ross had what he needed.

As he disappeared out the door I switched my attention to the all-important black phase. The supercell search that Nancy and I had started the previous evening was still in progress, but the incomplete diffractometer output told the story. One perovskite axis was tripled. The unit cell appeared to be a perovskite, with a shape like three cubes stacked on top of each other. The crystal symmetry was tetragonal, characterized by a unit cell with a 3.9-angstrom square base and 11.7-angstrom height.

I quickly reviewed the details in my mind: three large metals (one yttrium plus two bariums), three small metals (the coppers), and a tripled cell. It suddenly seemed so logical to me—the number three was the key. Put the three small copper atoms at the corners of the three cube-shaped subcells; alternate the three larger barium and yttrium atoms in a regular pattern: Ba-Y-Ba, Ba-Y-Ba in the tripled perovskite large-metal sites. It all made sense. The chemical composition matched the perovskite triple-cell structure perfectly.

I decided to start data collection immediately with the new tripled cell. I increased the measurement time for each reflection in the hopes of improving our counting statistics and thus seeing a few weak reflections that might otherwise be lost in the background. It was a trade-off, for longer collection times per reflection meant fewer reflections each day. The diffractometer was on its own as I dashed off to my fast-approaching Sunday afternoon Brahms concert.

Ross tackled the green phase with renewed determination. Solving a crystal structure means fitting atoms into a box or, as crystallographers call it, the unit cell. The green-phase formula is Y_2BaCuO_5, a total of nine atoms. From symmetry arguments Ross knew that he had to fit four of these chemical formula groups—a total of thirty-six atoms—into a 6 by 7 by-12-angstrom box.

It was like a complex three-dimensional puzzle, the solution of which was made only slightly easier by a few empirical rules of the game: atoms are usually at least 2 angstroms apart; several oxygen atoms surround each metal atom; metal-oxygen distances are ordinarily shorter than metal-metal or oxygen-oxygen separations. But even with such guidelines, there is usually no way to know in advance where any given atom might be found in an unknown crystal structure. A structure solution can take days or weeks, even with the most advanced computers.

Ross's first step was to rerun the Patterson maps that had failed the previous day. As he expected, the maps were much clearer, and the approximate position of barium atoms could be inferred. Even though the other atoms—the yttriums, coppers, and oxygens—were impossible to locate from the Patterson map, it was enough to know the barium positions. The nut had been cracked, for once those heavy atoms were located a complete solution was just a matter of time. Solving a crystal structure is something like making a jigsaw puzzle: as the picture is filled in it becomes easier to fit in the remaining pieces.

Ross began the next step of the green-phase study by describing the part of the structure that he knew: four barium atoms floating in a unit cell 6 by 7 by 12 angstroms in size. The rest of the box, which he knew held the other atoms, was empty in Angel's first model. Given this crude model, he was able to apply Fourier analysis, a crystallographic technique that converts observed x-ray intensities into "maps" of atom positions in the unit cell. Fourier methods are a powerful tool for solving crystal structures, but the technique cannot be applied unless at least one atom position is known. Fortunately, Ross had already pinpointed the barium atom positions, and as a result other atoms in his Fourier maps began to appear.

Almost immediately he spotted peaks in the maps corresponding to the yttrium atoms. Angel constructed a new, more complete, structure model. This time twelve atoms—four bariums and eight yttriums—were placed in the box.

From then on the structure solution was a straightforward, if repetitive, process. With more atoms in his model, Ross obtained sharper Fourier maps with more peaks. Copper atoms were located and added to the structure model, which in turn improved the resolution of the subsequent Fourier map. Next several oxygen atoms appeared. It took perhaps a dozen cycles, with one or two false atom positions along the way, but by early evening all thirty-six atoms in the box were found. Ross sat back in his chair and looked at the latest printout, and gradually it dawned on him that he had the solution.

The green-phase structure had been solved! It wasn't at all like a scene in a movie. There had been no sudden inspiration, no single moment of discovery. The green-phase structure appeared slowly and methodically out of the noise of Fourier analysis. Ross had oozed his way into the solution; it took the exhausted postdoc a moment or two to realize that he had really done it. After almost thirty-six straight hours of work Ross had discovered a complex atomic pattern in which each yttrium atom was surrounded by a triangular prism of six oxygens. These prisms shared edges to form infinite chains along one crystal axis. Copper and barium atoms, fitting neatly between the yttrium chains, provided the glue that held the chains together. Ross's solution was mathematically rigorous and positively elegant. And it was unlike any structure any of us had seen before.

An exultant Ross phoned Charlie Prewitt and me to spread

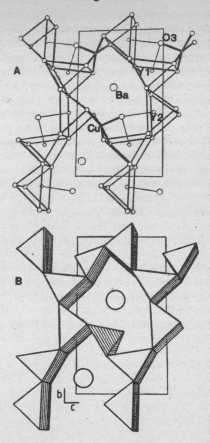

FIGURE 6. *The green-phase structure has chains of yttrium atoms, each of which is surrounded by seven oxygens. These yttrium chains are cross-linked by copper and barium atoms. The drawing on top (A) is a classic "ball-and-stick" representation, and the figure on the bottom (B) is a more stylized "polyhedral" representation.*

the good news. We were delighted and impressed with his remarkably fast solution to a complex structure. He then went home to share his success and a late dinner with Nancy, who was by this time getting a little tired of her mate's green-phase obsession. But, in spite of his euphoria, Ross's mind had not yet dropped the problem. As he and Nancy were talking over dinner he thought of one unresolved detail.

There are many different types of orthorhombic symmetry, depending on the number and types of mirrors and rotations that can be used to describe equivalent atom positions. Ross's new structure could fit equally well into two of those symmetry "space groups." To anyone looking at the structure drawing this subtle difference would not matter a whit. But in Ross Angel's mind the problem wasn't completely solved until he knew which space group was correct.

The answer could have kept until the morning, but Ross had to know. He trudged back to the Lab. There he compared the two possible space groups. For each space group Ross had to prepare a computer file with the appropriate symmetry relations, the list of atoms, and the observed x-ray intensities. Then he initiated a program that refined the positions of all the atoms.

Shortly after midnight he was sure. At last we knew the chemical composition and crystal structure of the green phase. Half of the superconductor problem was solved. Our pleasure and excitement at this achievement was only slightly dimmed by the fact that the superconductor was the black phase.

Chapter 7.

GIGO

Monday, 2 March 1987

The cat was out of the bag. Paul Chu's superconductor article hit the scientific world like a bombshell. The front page of the *New York Times*, a full-page spread in *Time* magazine, feature articles in *Science, Science News,* and virtually every other general science periodical—all lauded Chu's breakthrough. Several reports were laced with unreserved hyperbole and visions of a radically altered future society.

But even though the 90-Kelvin superconductor discovery belonged to Paul Chu's team and his team alone, the work of other groups on the yttrium compound was already becoming widely known. It was clear that Bell Labs, IBM, Berkeley, and Argonne, as well as the Beijing group, had all been hot on the trail for several days.

We had made progress. We had solved the green-phase problem, though the green compound was not the superconductor. We knew the composition and tripled perovskite unit cell of the black, superconducting phase, and we could make an educated guess as to its structure. But guesses weren't good enough. We needed good x-ray intensity data, and the black crystals on hand were lousy. We seemed to be stuck until Paul's new samples arrived.

My first priority of the week was to contact Art Freeman, whom I had promised to call with any new data. Although I knew the black phase was now his only interest, I decided to tell him of the green-phase solution. Perhaps, with progress on the black phase so slow, I wanted to prove to him that we could solve a structure. Or perhaps, with so many research rivals, I felt a need to touch base with a friendly collaborator. Unfortunately Art didn't make me feel any better. I reached him at his desk number, where he answered with a curt "Hello?"

I started with a casual greeting. "Hi, Art, how's it going?"

He wasn't in a chatty mood. "Listen, Bob. Have you heard about Bell Labs?" His question was loaded. Evidently Art had revelations far more interesting than my green-phase news.

"No, what about them?"

"Rumor has it they've solved the black phase. It's supposed to be a cubic perovskite." He was informing and probing me at the same time.

"How did they do it? Single crystal?" I was really curious now. And worried.

"No. Evidently it's powder work. I heard the story discussed over lunch at Argonne. So what do you think?"

I didn't even stop to think about his reference to the Argonne National Laboratory near Chicago. I was too concerned about the implications of the Bell Labs story. "Well, of course they're wrong. We know it's a triple-cell tetragonal perovskite. There's no way they're going to solve it from powder data."

"So, Bob," he replied, "what can you tell me?"

"I wish I had something new for you, Art. But we're stuck until we get better crystals from Paul. We think there's a Ba-Y-Ba ordering that gives rise to the tripled cell, and it makes sense to have coppers at the cube corners, but there's no way we're going to see the oxygens without better x-ray data. We *have* solved the green phase, though—last night," I added.

"Well, we have to have the *black* structure. Call me when you know more." He obviously had shifted his interest from the green phase, and I wasn't much help to him. But he added in a friendly closing, "Give my best to Margee and the kids."

Of course Art was right, but it was depressing, nonetheless. Having sought friendly reassurance, I came away from the conversation agitated and uncertain of our next move. Perhaps Nancy would have some new ideas. Over the weekend she had been giving the superconductor sample a thorough examination, looking for larger crystals and taking their x-ray photographs.

For more than a day one diffractometer had been dutifully amassing data on the first black crystal, the one that I had mounted more than a week earlier. The peaks weren't great but we had grown used to them. We were more concerned about two other problems that plagued the experiment. First, the intensities were so weak that long count times of about ten minutes per reflection were required just to see most peaks above the background level of scattered x-rays. With perhaps 2000 available reflections, it meant a long and tedious data collection.

But even more serious was the fact that the intensities we *did* collect might not mean very much. Barium and yttrium are very efficient absorbers of x-rays. Even our tiny, 40-micron black fragment absorbed more than half of all incoming or diffracted x-rays, and that absorption in turn altered the observed intensities. Under most circumstances, with a single crystal, we could have corrected for the effect using any one of a number of standard computer programs.

But we didn't have a single crystal. Our black-phase sample consisted of four interlocking crystals of similar size. Only one crystal diffracted the x-rays that we measured, but all four crystals absorbed x-rays. There was no way to correct for absorption, so all our intensities could be off by plus or minus 50 percent. We tried to improve the data by employing an "averaging" technique, but that procedure was tedious and time-consuming. Even worse, it was scientifically sloppy, because there was no way to be sure that our data were valid. Under normal circumstances we would never have used multiple crystals, but with time at such a premium we had no choice.

The South Huber diffractometer, which had the original black-phase crystal, continued to slog through reflections in the hopes that something useful might eventually result. But with the green phase solved, we had a second diffractometer, the North Picker, to devote to black-phase work. I found Nancy in the crystallography lab fiddling with the x-ray camera. She spoke first.

"I *think* I may have a couple of other crystals to try."

"Good! Are they any larger?"

"Well, the fragments are larger but I suspect they're multiple crystals again. In any case, the spots on the Polaroid look stronger." She showed me her latest film effort. The image was the best the superconductor had yet offered. Three intense white spots, along with four or five weak ones, contrasted with the black background.

"That's great! Let's put this one on the diffractometer. I assume that we can remove the green crystal for now?"

"Sure, Ross won't mind. He's still asleep, and I don't expect him soon."

We transferred the large black chip, a 60-micron oblong grain now dubbed crystal #2, to the now idle North Picker. The green crystal and its metal mount were carefully set aside for possible future study. We repeated the familiar routine of crystal alignment, reflection location, and centering.

It took the better part of a day for us to get the bad news. Crystal #2 was a dud. The reflections were stronger, to be sure—as many as 400 x-ray counts per second. But the maxima were broad, split, and horrendously smeared. We couldn't even get a decent unit cell. We were faced with a double difficulty. If, after many hours of searching, that was the best-looking crystal, then we would never solve the structure. And even if we could get a good set of intensity data, the peak splitting and smearing might be symptomatic of a structural distortion—called twinning—that we might never unravel. Our only recourse was to try Nancy's other "good" crystal.

Black crystal #3 took center stage at 3:30 that afternoon. It wasn't great, but it was better than #2. The intensities and peak shapes seemed to be about the same as those of the first crystal, so we went ahead and started the familiar sequence of steps leading to refinement of the unit cell. By 5:30 P.M. the diffractometer was on its own. There was nothing to do but wait.

Tuesday, 3 March 1987

With two diffractometers cranking away, and at least another day before we had enough intensity data to make even an initial stab at atom positions, it was a good day to get caught up on neglected reading and writing.

The phone was strangely quiet. Paul and Art knew that we were unlikely to make any significant discoveries without the new sample, so we didn't have any reason to speak. No one else called all day either. Holed up in my office, I felt isolated. The silence, after days of activity and pressure, was distracting, almost eerie. I tried to think about the unknown superconductor structure, with its tripled perovskite unit cell and YBa_2Cu_3 metal composition. I even sketched out a few plausible oxygen arrangements. But mostly I frittered away the time, unable to concentrate on anything.

Dave Mao was more productive. He was about to fill in another gap in our description of the superconductor. Early in

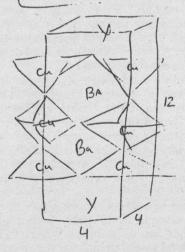

FIGURE 7. *By March 3 we had determined the basic elements of the structure. This sketch, scribbled on the back of some discarded computer output, included most of the important features. The placement of yttrium atoms at the ends of the unit cell rather than the center is a purely arbitrary decision; the structure is the same either way.*

the morning he traveled to the sprawling Reston, Virginia, headquarters of the United States Geological Survey, where the density measurements were to be made.

We already knew the chemical compositions of the two phases, as well as the size and shape of the unit cells. But we had only guessed the number of chemical formula groups in each unit cell. Ross, for example, had assumed on the basis of symmetry and his knowledge of structural systematics that exactly four units of composition Y_2BaCuO_5 filled each 6-by-7-by-12-angstrom unit cell of the green phase. Similarly, we strongly suspected that each 3.9-by-3.9-by-11.7-angstrom unit cell of the black phase contained one yttrium, two bariums, and three coppers. But not all compounds are that simple. Some materials, especially those with odd electrical properties (like superconductivity), are known to have fractional numbers of metal atoms. So instead of exactly three coppers or two bariums per unit cell, it is possible to have fewer atoms in a "defect structure." We wanted to be sure.

Density is simply mass per unit volume. Solids with heavy atoms such as lead, or with closely packed atoms as in steel, are correspondingly dense. A precise determination of density can reveal the number of atoms in each unit cell.

Under most circumstances, when reasonably large samples are available, density is easy to measure. But the black and green compounds that made up Chu's superconductor were microscopic. Dave Mao had spent hours isolating enough green and black bits to attempt micropycnometry, the most sensitive technique to determine density developed to date. John Marinenko of the United States Geological Survey had one of the few micropycnometers in the country, so Mao took a trip to Reston.

The technique is simple in concept. Weigh a tiny vial of liquid of known volume, place the solid sample in the liquid, and measure the new weight and the displaced volume of liquid. With properly calibrated glassware, microscopic samples can be measured. Making the arrangements that Tuesday was tedious, and it took the better part of a day to prepare the liquid mounts and repeat each run. As expected, both phases had densities about six times that of water. Another piece of the superconductor puzzle was in place.

Wednesday, 4 March 1987

Time was running out. We had to start analyzing the black-phase data. After four days of continuous diffractometer operation we had about 500 reflections from crystal #1, representing perhaps thirty-five symmetrically different reflections that were strong enough to see. We needed at least twice that much data to do a complete structure refinement, but it was enough to get started.

Each day I had spent thirty minutes or so evaluating the hundred-plus peaks from the previous twenty-four hours. That being done for the latest block of data, Ross Angel, Nancy Ross, and I assembled in Larry's office where we began the structure analysis. We had to fit about thirteen atoms—one yttrium, two bariums, three coppers, and six or seven oxygens—into a 3.9-by-3.9-by-11.7-angstrom box. It should have been an easy job. After all, we knew that the structure was related to perovskite and there weren't very many atoms to locate.

We began just as Ross had with the green phase, by calculating a Patterson map. The map was certainly pretty, with whopping big peaks at regularly spaced intervals. But the trouble with perovskites is that *all* perovskite Patterson maps look pretty much the same. All peaks in a Patterson map correspond to the interatomic distances and directions in the crystal. In a perovskite those distances and directions always correspond to a cube edge or diagonal. All we could say was that there was nothing that was obviously *not* a perovskite-like atom position.

Common sense, rather than any insight from the Patterson map, guided us from there. We had to build a perovskite from the few available parts—yttrium, barium, copper, and oxygen atoms. It was a three-dimensional jigsaw puzzle that had only a few plausible solutions. We had to locate six metal atoms; a triple-cell perovskite has six metal sites. It seemed more logical to place our six metal atoms (one yttrium, two bariums, and three coppers) on the six available metal positions, rather than invent a new site somewhere else in the box. It was also logical to assign the larger Y and Ba metals to occupy the larger central perovskite sites, while the small Cu atoms filled the smaller perovskite corner sites. That was the structure we all expected. But the x-ray data would not cooperate.

We tried this logical arrangement of metal atoms, and even without incorporating any oxygens into the model we got reasonably good results. The "residual," a somewhat arbitrary number that told us how well our observed data agreed with the calculated structure, was down to 20 percent, a value that would usually signify a correct basic structure. But virtually *every* model we tried gave just about the same 20 percent residual as long as metal atoms occupied perovskite positions. We could swap copper for yttrium or yttrium for barium and it didn't make any difference to the residual. With our mediocre data we could not prove the suspected metal positions, much less guess at the locations of the weak-scattering oxygens. We had to have more and better data.

When Art Freeman called late that afternoon I could only reiterate our strong hunch regarding the Y, Ba, and Cu positions. I could sense his frustration at having to wait for our structure solution, but I urged him to hold off his expensive calculations for another couple of days while we fought for improved results.

Thursday, 5 March 1987

We were very close. We knew composition and density for both green and black phases, we had solved the green-phase atomic structure, and we had the unit cell of the perovskite-like black phase. With Paul Chu's visit just a few hours away it was time for me to consolidate our findings.

Downstairs in my office I typed a one-page summary, while upstairs two diffractometers continued to collect black-phase data and Nancy, Ross, and Larry continued to puzzle over the structure. With another day's worth of reflections the metal atom arrangement was becoming better resolved. But it was becoming increasingly obvious that, while we could deduce the basic metal arrangement that led to the tripled perovskite cell, the exact oxygen positions would remain elusive until better crystals became available. We could only hope that Paul's new sample, which he promised to bring that afternoon, would do the trick.

With everything under control and little to do on the super-conductor but wait for Chu, I actually managed to clear a path down the middle of the papers on my desk. I even joined the half-dozen lunchtime volleyballers, who donned shorts and T-shirts in defiance of the gusty 50-degree day. Exercise, I rationalized, is a great way to clear out the mental cobwebs.

Paul Chu arrived at my office punctually at 2:00 P.M., escorted by Dave Mao. There was nothing extraordinary about him, though seldom have I met a scientist who seemed at once so naturally warm, unassuming, and soft-spoken. With his boyishly unruly black hair and blunt-cut bangs he appeared much younger than his forty-three years. He resembled Dave in his slender build and modest height, several inches shy of six feet. In fact, standing side by side the two men could easily have been mistaken for brothers.

I invited them to sit down, which they did with a charming mixture of Oriental politeness and Western enthusiasm, and I called upstairs for the other members of the superconductor team to join us. As we waited for the crystallographers Paul broke the ice.

"I'm very pleased that you agreed to help me on this project. It's really great."

"We're the ones who are grateful," I said. "Your supercon-ductor discovery is amazing."

We continued to exchange compliments until Larry, Rus, and the Rosses arrived. I made the appropriate introductions and we all stood around my cramped quarters for a moment or two shaking hands and smiling at one another. Chu seemed relaxed as he greeted each of us and his manner was unaf-fected and kind. Nevertheless, all of us, except perhaps Dave, were a bit nervous and somewhat in awe of the man. But soon his enthusiastic manner and ready smile put us all at ease. Dave, Ross, Nancy, Rus, and I each in turn showed him our results and he was once again generous in his thanks and compliments. He almost made us feel as if his contribution was trivial while our work was historic; that we were doing *him* an extraordinary favor by working on the superconductor. While none of us was about to adopt this view, we enjoyed his appreciation for our efforts.

Following our show-and-tell, I gave Paul the summary

sheet with all the current data on the two phases. In return, we received the latest and best superconducting sample—nearly pure black phase. He was confident that this specimen would be better because larger crystals had been clearly visible to him in the microscope. We were delighted and promised to get to work immediately. Eager to begin, the other scientists headed for the x-ray lab, leaving me alone with Paul.

Given the excitement, almost euphoria, of the visit, I was totally unprepared for Paul's next remarks and change of mood. He sat down, looked me in the eye, and adopted a quiet, sober tone in marked contrast to his ebullience of a few moments before.

"Bob, I know you've been in touch with Art Freeman. How much have you told him?"

The way he asked the question conveyed more than the words themselves. He was asking out of more than mere curiosity.

"I just talked to him last night. I told him everything that I told you. I thought it was OK. Aren't you working with him?" My tone was almost pleading.

"I'm sorry, Bob. Art has evidently been collaborating with Argonne—he may have given them your structure data. Also, I was told that he has filed a patent for a superconductor formula related to mine."

I was stunned. Art Freeman had betrayed us? It seemed impossible. He had been intense and aggressive at times, but always sincere and always driven, or so I thought, by the simple desire for knowledge. Why would he do it? Was it simply a clumsy effort to seize a portion of the superconductor bonanza? Did he use our data to barter for the results from the Argonne superconductor team? Or perhaps his first loyalty was really to his Illinois-based coworkers. Whatever the case, the worst had evidently happened. All of our hard-won structure and composition data were apparently in the hands of rival workers. Now the race was dead even.

My first reaction of dismay turned to embarrassment, for I was the one who had passed on the vital information. How could I have been so naive? The superconductor secrecy, so unlike anything I had ever done before, was alien to me. Paul had every reason to blame me, for he had much more to lose than I, but he sensed my distress and was consoling in his words. "It's OK, Bob. You couldn't have known. I trusted him too."

Paul continued, "But from now on, let's keep everything confidential. Let's just say we'll release the data at the March eighteenth meeting. In the meantime I'll get a preliminary paper off to *PRL* to establish priority on the structure work."

As he spoke, my mood shifted from self-condemnation to anger toward Freeman, the alleged deceiver. I, too, could deceive. Art would undoubtedly phone again, and I could imagine any number of plausible—but totally wrong—structures to feed him. Let him spend his fortune in computing on a hoax. Let him stand up in front of the American Physical Society and make a fool of himself. I resolved to pay back Art Freeman in full.

Paul quickly took his leave, having accomplished his mission, and headed to the next stop in his hectic schedule. My good-byes were sincere but distracted, for my mind was in turmoil. I burned to get even.

But revenge would have to wait. There were new samples to mount and the structure was still ours to solve, if only we were swift.

Ross and Nancy had already selected several fragments of black phase that were as long as 40 or 50 microns. We chose the two best, dubbed them crystals #4 and #5, and mounted them on needles with quick-drying glue. One would occupy the North Picker diffractometer, replacing crystal #3, which was unlikely to yield the secret of the oxygen positions. The other would inhabit Ross Angel's Picker Number Two, which he gladly relinquished to the chase. All three automated diffractometers would work together. Our hard-won experience from the earlier sample enabled us to photograph, center, and orient the two promising black bits in less than an hour. Both superconducting specks gave reasonably strong diffraction, and one of them, crystal #5, even seemed to be a single individual rather than a multiple crystal. We would keep our fingers crossed during the night as reflections were centered and unit cells refined, prior to the critical data collections. We all sensed that this was our last chance to solve the superconductor riddle.

Friday, 6 March 1987

Crystal #5 was the best so far. It appeared to be almost entirely one individual, with only minor x-ray diffraction from a second tiny crystallite. The peak intensity was also an improvement, with the strongest reflections producing more than 300 x-ray counts per second. That was about twice the intensity of crystal #1, so data could be collected much faster. The unit-cell refinement, completed during the night, was also the best so far.

When Nancy, Ross, Larry, and I conferred at 9:00 that Friday morning, our decision was unanimous: begin data collection on crystal #5 as quickly as possible. We anticipated a collection rate of at least 350 reflections per day; we hoped to have sufficient data to solve the structure on Sunday.

In the meantime, crystal #1 grudgingly yielded its tantalizing trickle of new intensities to play with. Under most circumstances we would have deferred all attempts at a structure solution until a complete data set was available. But in the interest of speed we resumed work on the suspect data set from crystal #1 in the hopes that it would point us in the right direction until Sunday.

As we were talking my call bell rang. The Geophysical Lab building retains several vestiges of its early twentieth-century origin. One of the least endearing of these traits is an archaic telephone system featuring numbered call bells that resound throughout the building and grounds. Twenty-four was my signal. I excused myself, ran downstairs to my office, and answered. It was Paul Chu.

"Hello, Bob. I hope I'm not interrupting you?" he asked.

I hadn't expected to hear from Paul so soon, having just seen him the previous afternoon. I wondered if there had been some new development.

"Hi, Paul. We were just working on the new sample. It looks good. What's up?"

"I just wanted to tell you not to worry about Freeman. I know you feel badly. You didn't do anything wrong; just don't let it bother you." He sounded almost fatherly in his concern.

I wasn't sure how to respond. "Thanks, Paul. I really appreciate that. It does bother me. But you can bet I won't give out any more information."

"That's fine, Bob. I'm sure you'll get the structure. Good luck."

With that he closed the strange, short call. And while I wasn't completely sure why he had gone to the trouble, I felt better for it. He had somehow defused my anger toward Art Freeman and set my sights more squarely on the black phase. In my initial fury I had wanted to strike back at Art. But, in spite of my emotions, I had quickly realized that I could never give out false data. Such an action would be contrary to everything that science had taught me. Withholding information was a different matter—Freeman would gain nothing more from me.

I hurried back upstairs, intent on doing as much as possible to expedite the process. The latest twenty-four hours' worth of data from crystal #1 had to be processed, so I dismounted the magnetic tape and evaluated those reflections. The newly revised and consolidated data set had more than seventy observed reflections, enough to be fairly certain about the approximate metal atom positions. I joined the Rosses and Larry, who spent the better part of the morning trying various configurations of yttrium, barium, and copper atoms. Our preferred metal arrangement with coppers at cube corners and barium and yttrium at cube centers still clearly seemed to be the best, yielding a residual of about 15 percent without including the oxygens.

By the afternoon we were ready to look for oxygens. We employed the same Fourier process that Ross had used on the green phase. We hoped that by knowing all of the metal atom positions the unknown oxygen locations would appear on our Fourier maps. Larry started the calculation and we watched the computer terminal expectantly. The resulting map was a mess. There were positive features, to be sure, but they were hardly the neat little spherical forms that we had hoped to see. Rather, the Fourier map showed broad bands of positive intensity separated by physically impossible negative regions. The data from crystal #1 just weren't good enough.

In desperation, and against Larry's better judgment, we

tried a logical model with both metals and oxygens in the normal perovskite positions. This model was "refined" by letting the computer program automatically adjust the starting model to find the best fit with the data. We not only let the program vary the oxygen positions, but we also adjusted their occupancies—the fraction of each site actually filled by oxygen. Since we knew that about one-third of the perovskite oxygens were missing, this procedure seemed a good way to find out which were which.

At first the results looked terrific. The refined model had a residual of only 6.5 percent, a value better than many published structures. But a closer inspection of the computer output dampened our enthusiasm. Some oxygen positions had disappeared as we had hoped, but others showed occupancies of almost 2.0—a physically impossible situation with two oxygens sitting on top of each other. (Occupancies, averaged over a whole crystal, have to be between zero and one.) Furthermore, the resultant bond distances were not logical. One copper-oxygen distance was only 1.3 angstroms, far too short for any stable structure.

After examining the results in silence for several minutes, Larry opened his desk drawer and took out his thickest-tipped magic marker. Without a word he scrawled across the printout:

GIGO!

Nancy started laughing but I was in the dark.

"What's so funny? Do you have a new idea, Larry?" I asked.

He looked surprised. "You mean you've never heard of GIGO?" He raised his eyebrows in mock contempt. "Garbage in—garbage out! We ain't gonna do nothin' with this data set."

I had to agree. We had gone about as far as we were going to go with poor little multiple crystal #1. "Looks like we'll have to wait for number 5 this weekend," I acquiesced.

"I'm afraid I won't be coming in, Bob." Larry sounded depressed. "My blood pressure has shot back up. It's all this superconductor business."

For more than a year Larry's blood pressure had been under control, but the damn superconductor was putting us all under stress. It was a terrible time to lose him, but this recurrence of the condition was not to be ignored. "I'm sorry, Larry. I wish there was something I could do."

"I'll be OK. Hey, I might even try to do some computing at home on my system. I'll need to know the name you give to the new data file."

"How about BLACK5.DAT? I'll use something like that."

"Good, thanks. I'll keep in touch." As our group split up for the evening I heard the old familiar, aggravating call bell: ding-ding, ding-ding-ding-ding. Another phone call. I checked with our operator.

"Dr. Hazen? Professor Arthur Freeman on line four."

That was one call I wasn't about to take. "Please tell him I'm not available to take any calls. Thanks!"

Chapter 8.

1-2-3 Superconductor!

Saturday, 7 March 1987

The new data from crystal #5 were pouring in. By 10:00 A.M. I had processed more than 300 reflections. The maxima were generally broad and a few even had slightly doubled peaks, but overall the reflections were much stronger than from crystal #1. Furthermore, it looked as if we could actually make a reasonable correction for the considerable effect of x-ray absorption by the crystal. I labeled the new data set BLACK5.DAT, as promised, and let the Rosses go to work on the oxygen search. Meanwhile, as much out of habit as need, I looked at the next batch of reflections from crystal #1. It was probably a futile exercise, but I couldn't quite admit that we had wasted so many days on a useless crystal.

There is more to solving the perovskite structure than just locating atoms. The basic atomic configuration that we had inferred could distort ever so slightly into any one of a half-dozen symmetries, called "space groups." The difficulty in locating the oxygen atoms was compounded by the fact that if we used the wrong space group, spurious atoms might pop up in the wrong places. There were literally dozens of possible models to try, and superficially they all looked very much alike.

There were still not enough data from crystal #5 to gain the resolution we needed, but a few things were gradually becoming clear from the partial set. After many electron density maps, continually upgraded every few hours as new #5 data were processed, a few of the oxygen positions were becoming more obvious. In a true perovskite there are oxygen atoms halfway between every pair of small metal atoms—the copper atoms in our perovskite. We found that oxygens were present between pairs of coppers in the planes just above and just below the yttrium positions. Those oxygen sites appeared to be fully occupied. But at the yttrium level itself all of the oxygen atoms were gone. There was no evidence for oxygen electron density at those sites.

Like video game addicts we kept trying models and making maps. I processed the new data every hour now, only twelve or fifteen new reflections at a time. We were sure that the solution would soon be ours.

I had promised Margee that I would meet her and the kids for a pizza dinner out. My midafternoon estimate was to meet at 5:00 P.M., but at 4:30, with several different models yet to try, I knew I'd never make it. I called home briefly to adjust the schedule to 5:30 and went back to the computer.

We shuffled oxygens in the remaining possible sites, at the barium levels and at the copper level between the two barium atoms. All of the possible structures gave similar residuals of about 8 percent, and all had flaws. It was 5:00 P.M. and we still had more to do.

My second call home was not unexpected. "How about six?" I asked, without explanation.

"Fine, no problem. See you then." It may have been our shortest phone conversation on record.

The family's volatile eating schedule was thrown off once again by a particularly knotty problem in describing one of the low-symmetry variants of the perovskite structure. The lower the symmetry, the greater the number of different atoms that must be described separately to build the unit cell. It was almost 6:00 P.M. when we got it right.

One more time. "Hi! Guess what? It'll be six-thirty for sure. I promise."

I actually made it to the eating place by 6:50, where the large pizza with extra cheese and sausage was just being sliced. Other than her lightly mocking "Nice timing, Bob," Margee seemed unperturbed by my erratic schedule. Perhaps she sensed that we had almost solved the problem and that things would soon be back to normal.

"By the way, you got a call from Art Freeman just before we left home." She knew the whole Freeman saga, and I could just imagine the icy tone that she must have used.

"He actually called at home again? What did you say?" I asked.

"I just said that you weren't available and that I didn't know how late you'd be. I also said I was going out this evening. Did I do OK?"

"Well, you didn't lie, and I'm sure not going to return his call."

Sunday, 8 March 1987

For the second Sunday in a row it was a glorious day, with brilliant sun, deep blue skies, and temperatures in the seventies. It was as if nature, wishing to hold onto her secrets for a while longer, was trying to lure us away from the Lab. But we were not tempted. The black-phase structure was within our grasp. We knew the composition, we knew the cell unit, we knew the metal atom positions, and now we had the data to find the missing oxygens.

Crystal #5 data collection was completed at 10:00 A.M., and I immediately set to work processing the final reflections and adding them to the list. By 11:00 the BLACK5.DAT file boasted more than 600 reflections. Ross and I were at Larry's office terminal, ready to refine the most promising models, while Nancy Ross made additional calculations in her office. A quick scan of other activity on the Lab computer system revealed that Finger was also busy, logged on at his home computer terminal and working on the problem. It was an all-out assault—and, for the first time, we had the data necessary to solve the structure.

Ross pounded the keys of the computer terminal, editing the structure model file for perhaps the hundredth time. "Which model shall we try first?" he asked me. "Shall we go for centric or not?"

That was a key question. In centric crystals there exist special points within each unit cell, points that have high symmetry. Any atom on one side of such a "center of symmetry" will always have a second atom lying precisely on the opposite side of the center. The number and location of these highly symmetric positions can have a profound effect on the electronic properties of the material. Even a slight deviation from centric symmetry can result in a "polar" crystal, that is, a crystal having positive and negative ends in an electric field. Such polar crystals are essential components of microphones, speakers, and dozens of other products worth billions of dollars annually to the electronics industry.

In some perovskites every atom is at a center of symmetry, but in many others one or more atoms is off center. Given the extraordinary electrical properties of the superconductor, we wanted to resolve this question. The problem was that all of our potential structures, centric or not, looked just about the same. There was no guarantee that we could ever resolve the question.

"Let's try the simplest thing first. Let's do four over em em em." My vote was for the highest-symmetry, centric structure. In that model the yttrium atom was at a symmetry center, so the two barium atoms, which lay on opposite sides of yttrium, were symmetrically equal, as were groups of copper and oxygen atoms. Not only was the model the simplest to describe, but it was, to my mind, the most pleasing aesthetically. All of the oxygens were constrained to be in perfect planes, the coppers perfectly aligned. It was a beautiful structure.

It didn't work. The model refined to about 9 percent, but there were problems both with unreasonably short bond distances and with impossible oxygen occupancies. We had to consider lower symmetries.

We talked about the problem for a bit and started on the opposite tack. Rather than imposing the highest possible symmetry on the structure, thus perhaps forcing the atoms into an unhappy configuration, we chose a model with the lowest possible symmetry consistent with the x-ray data. In that way more atoms were free to shift into comfortable positions. We included the three coppers, two bariums, and the yttrium, as well as the four oxygen atoms that we thought were adjacent to the yttrium layer. The remaining oxygens were omitted for this attempt. If we were right then the model should refine properly and give a lower residual. Then we could use that known structural block to help find the rest of the oxygens. If we were wrong, then the atoms and occupancies might shift into physically impossible values—a process we aptly called "blowing up."

This time we were on to something. The residual plunged to below 6 percent, and all of the atoms—for a pleasant change—were reasonably well behaved. Occupancies were all close to 1.0, and the metal-to-oxygen bond distances were all in the normal range. All the numbers fit!

"Let's call Nancy and draw this," I said, certain that we had the answer. "I can smell this structure—it's really there." Ross called Nancy down to Larry's office while I set furiously to drawing the suspect structure.

All that was missing was the position of the remaining two-plus-a-fraction oxygens. And there were only three perovskite-like oxygen sites left to try: two positions at the levels of the two barium atoms, and one position in layers between the barium levels. An electron density map was the next logical step. With Nancy and me looking on, Ross ran the program. Clear features appeared at all of the three possible oxygen sites. We recorded the apparent coordinates and reran the structure refinement, this time with all the atoms.

It worked! A residual of only 4.9 percent. All of the atoms were well behaved. The occupancies of the three new oxygens were all fractional, indicating to us that all three sites, averaged over the entire crystal, were only partially filled. On a random basis some unit cells had the oxygens and others didn't. The errors on those occupancies were very large, close to plus-or-minus half an oxygen in uncertainty. But that result was not too surprising, given the weak x-ray scattering by those unfilled sites compared to heavy metals. One 1.7-angstrom copper-oxygen distance seemed a bit too short; we were expecting distances closer to 2.0 angstroms. But we decided that large bond distance errors could account for the discrepancy.

There was no question in our minds. After more than a week of flirting with the correct solution we had finally solved the structure of the superconductor. A layered arrangement—that was the key! No other structure has these strange and beautiful layers of copper and oxygen atoms, in plane after plane, separated by yttrium and barium atoms. We were sure

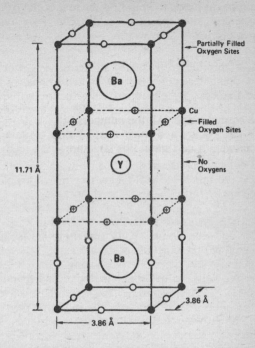

FIGURE 8. *Our first complete black-phase solution included several partially filled oxygen sites (open circles) near the barium atoms.*

that those unique atomic planes held the key to the origins of its superconductivity. What exhilaration, to know something that no other human being has ever known, something as fundamental as how atoms fit together to make reality. Such moments are few and therefore precious in a scientist's life.

Perhaps Larry had been monitoring our progress at home, or perhaps it was just a coincidence, but an electronic mail message came in even as were celebrating. Larry's home phone was busy providing a computer link to the Lab, but we could exchange information just by typing the message and using the computer's internal mail capability. We immediately sent him the good news:

WE SOLVED THE STRUCTURE! 4.9% SOLUTION.
EVERYTHING MAKES SENSE. CHECK FILE
BLACK5.PRM.

Larry, comfortably ensconced in his home's sophisticated
basement computer center, was delighted to hear of our success. He immediately examined the structure parameter file
that we had created and compared the results to his own. During the afternoon he had independently taken our low-symmetry approach even farther, eliminating even the basic
tetragonal symmetry and treating every atom separately. It had
been a tedious operation that hadn't worked well, but he
didn't seem to mind. He was happy to write down our results
and agreed to confirm them.

The news of our success spread rapidly. Within a few minutes
Dave Mao and Rus Hemley had come up to the lab and were
eagerly soaking up the details of the structure. Rus was particularly excited about the obvious layering. He had just begun
spectroscopic work on the superconductor and found that the
spectra were highly anisotropic, that is, they had a very different character in different directions. Prominent spectral
features, present when looking along the tripled axis, were not
present in other directions and vice versa. Such a result was
not expected from normal perovskites, but we had shown that
the atomic bonding was, indeed, very different perpendicular
and parallel to the copper-oxygen layers.

The Rosses still had some minor details to clean up, and I
asked them to prepare the tables of the refined atom positions
and interatomic distances. In the meantime I dashed downstairs to call Paul Chu and give him the good news. The time
was 5:00 P.M.

His laboratory phone was busy. Under most circumstances
I might have waited a few minutes, reading an article or writing a letter, but I was too anxious. I had to share our discovery. I redialed the number over and over. Finally Paul
answered.

"Hi, Paul, it's Bob. I've got good news."

He was excited. "Did you solve the structure?"

"Yes, it's amazing! It's a copper-oxygen layer structure.
All of the coppers are in square-planar coordination." I pro-

ceeded to recite the sequence of atoms and positions. "We plan to write the paper up immediately and send it off to *PRL* tomorrow. Of course you'll be a coauthor."

"Thank you, Bob. Please be sure to include my coworkers R. L. Meng and P. H. Hor. You should list me as last author, please."

I thought Paul's request a generous one and told him so. But it was time to get on with the serious business.

I was in an unusual situation. Knowing the crystal structure, I could easily predict any number of compositional variants based on Paul's original 1-2-3 ratio of yttrium to barium to copper. Some of those variants might work even better than the original material. We could have withheld the information and attempted to produce superconductors and submit patents at the Geophysical Laboratory for our own profit and glory. But that wasn't our style. So I told Paul my ideas.

"Paul, there are a couple of obvious chemical substitutions to try. Clearly an important feature is the square-planar copper. You should try substitutions of other square-planar metals, like iron or chromium or even platinum."

He was quick to respond. "OK, Bob, we'll try those."

"If nothing else it will help us to understand the mechanism of superconductivity. If, for example, you substitute a small amount of iron for copper or platinum for copper, those substitutions should result in a stable structure but they'll probably affect the temperature of superconductivity.

"Also, I'm very interested to see what pressure does," I added.

"We tried that some time ago. The critical temperature doesn't change at all—it doesn't work." Clearly he was way ahead of me on that one.

"Well, you know the obvious substitutions to try for barium and yttrium." Paul was well aware that strontium, calcium, and a number of other elements might substitute in the barium positions, while scandium and any of the rare-earth elements might proxy for yttrium.

"Yes, we've already made several other compositions that show the same high superconducting transition." That was another revelation that would stir up the physics community.

For a time we conversed about what the Geophysical Lab should do next. I repeated our request for more and better crystals, particularly of any other yttrium-barium-copper oxide phases that his group might stumble upon. With two new structures in a week, it seemed likely that the chemical system containing Y-Ba-Cu oxides was full of surprises.

Paul also suggested that I try to present our results at a special session of the American Physical Society on high-temperature superconductivity, to be held on March 18. He didn't have all the details, but he gave me the name of an organizer, Brian Maple, in California. I promised to call as soon as the paper was done.

Finally I brought up the painful subject of Art Freeman. "You know, Paul, Art has been phoning me every single day. I haven't answered the calls. Do you really think he's working with the Argonne group?"

For the first time that day Paul's voice was subdued. "I'm afraid so. That's what I've been told."

But I wasn't worried. "Unless they get single crystals and really do fast work I don't think they'll be able to crack this one right away," I told him. "It really took a lot of good intensity data. My guess is you'll have at least a week's head start in trying the new compositions."

"I'm sure you're right, Bob. Thanks again for all your help."

I assured Paul that we were the ones to do the thanking. It had been quite an adventure.

Chapter 9.

PRL

Monday, 9 March 1987

Many of my best ideas—the design for a complex experiment, the explanation for a puzzling piece of data, or even the text for a difficult manuscript—come to me unsolicited during morning showers. Immersion in the coursing hot water and billowing mists seems to open pores to my subconscious mind, which evidently doesn't get much rest at night. Nothing too dramatic came to mind that morning, but my day's responsibilities did seem remarkably obvious. It was time to write up our new results.

I gathered the gang of six at 9:00 that morning to mobilize the writing effort. We agreed that *Physical Review Letters*, which had published Paul Chu's original paper, was the best forum for our work, and we decided to be ready for the late afternoon overnight delivery deadline. Everyone would have to contribute something, and do it fast. Dave Mao agreed to summarize the all-important composition and density data. For many readers who wanted to synthesize pure superconducting material the composition information would be the most critical section. Charlie Prewitt offered to tabulate powder x-ray data, while Ross Angel described the green phase. Nancy would consolidate our new structure results on the superconducting perovskite and she also proposed to conduct a careful literature search for related phases that we might have missed in our hasty bibliographic work. Larry Finger volunteered to create a computer drawing of the new crystal structure, and I set about writing the black-phase text as well as the title, abstract, introduction, and conclusions that would frame the actual data.

Dave's job was probably the most straightforward because electron microprobe data were reduced to simple ratios of yttrium to barium to copper. His discovery of the integral proportions—2:1:1 for the green phase and 1:2:3 for the black phase—made things particularly simple. It was necessary to include experimental details such as the method of sample preparation, the type of microprobe, and the chemical reference standards used to make the quantitative analysis. Our analytical methods could not tell us exactly how much oxygen was present; thus we had to present a variable black-phase formula: $YBa_2Cu_3O_{6+x}$, where x was about 0.5.

Dave's report of densities—a single number for each phase representing the weight per unit volume—was also easy to document. Both phases were almost six times denser than water, the standard reference material.

Charlie Prewitt elected to present the powder x-ray information in tabular form, with only a brief paragraph describing the details of data collection. Powder data confirmed that only two phases accounted for virtually all of Chu's first superconductor sample—all but two of the more than forty powder peaks corresponded to diffraction effects observed in one of the two single crystals. Those remaining two peaks were very weak and impossible to characterize in greater detail.

The main body of the paper consisted of the two structure reports. Ross Angel had drafted a description of the green phase and produced an elegant structure drawing several days earlier. In that first draft he had proudly stated his findings:

> The structure of the green phase reported by Wu et al. has been determined by single-crystal x-ray diffraction methods. The new structure is orthorhombic with cell dimensions . . .

But we should have done our homework. The pace of work had been so accelerated that shortcuts had been taken. After discovering the green-phase composition we had made a hasty check of standard reference works, but we had neglected to conduct a more systematic and thorough literature search. In short, we blew it.

Science done in haste often exacts a price, and Ross Angel paid up that day. Nancy, who had agreed to scour the library for articles on yttrium-copper compounds, gave him the bad news; no words were spoken and none were needed. The article from a 1982 issue of *Journal of Solid State Chemistry* told its own story. The green phase had been described four years earlier.

Paul Chu was the first to discover superconductivity in the Y-Ba-Cu-O system, but he was not the first to work on that chemical system. In the early 1980s a group of French chemists in Caen, led by Claude Michel, synthesized and described dozens of new phases in copper-bearing systems. We found one article by Michel and Raveau describing the green phase, Y_2BaCuO_5, and a second paper by Er-Rakho, Michel, and others on a variant of the black-phase structure with the element lanthanum instead of yttrium. Much of their data had not yet made its way into *Structure Reports*, but we should have checked in other publications. It was small consolation to Ross Angel that his calculations corroborated and amplified the earlier work.

So now Ross's job consisted mainly of shortening his prose. Sadly, all he could do was confirm the findings of the earlier study rather than announce a new structure type:

> The green phase has the orthorhombic A_2BaCuO_5 structure, which was described on the basis of powder diffraction by Michel and Raveau. We confirmed the structure of the green phase....

We all felt sorry for Ross. It wasn't just the time he wasted—he truly deserved recognition for a first-rate piece of research and now there would be none.

Nancy's tables of the black-phase structure parameters were the heart of the manuscript. She prepared two lists, one with position coordinates for each type of atom, and one for distances between neighboring atoms. These tables would be amplified by the text, which included all the requisite crystallographic jargon.

Larry's task—to produce a clear and accurate drawing of the superconductor crystal structure—was in many ways the most difficult. He chose to implement a classic program ORTEP (for Oak Ridge Thermal Ellipsoid Plotting program). This versatile tool required a complex numerical input with coded information on crystal symmetry, atom positions, atom labels, interatomic bonding, and many other variables that are essential to an effective visual aid. Even though the atom positions were dictated by the refinement, Larry had to make a number of arbitrary decisions. He chose to employ a "ball-and-stick" drawing, in which the atoms appear as little spheres, with lines connecting metal atoms to the adjacent oxygens.

Once Larry defined the array of atoms to be drawn he used the program to rotate the image left or right, up and down, to find the most revealing perspective. This "best" image was further rotated 3 degrees left and then 3 degrees right to produce a three-dimensional, stereographic pair of drawings.

As the others tackled their chores I sat down at my computer terminal and began to process words. The title was easy: "Crystallographic description of phases in the Y-Ba-Cu-O superconductor." It wasn't fancy, but it got the job done.

The list of authors, which appeared directly under the title, wasn't so easy. I counted nine coauthors—the six of us from the Geophysical Laboratory as well as Paul Chu and his associates, Pei-Herng Hor and Ru-Ling Meng. The appropriate sequence of authors was anything but obvious. The Geophysical Lab workers would come first and the Houston synthesis team second; on that much we all concurred. Paul Chu modestly insisted on appearing last, though everyone who saw the article would know his real contribution. But the six Carnegie scientists had all worked hard for this paper and there was no obvious order. A simple alphabetical listing would have placed Ross Angel first, perhaps justly so. He had worked as hard as anyone and had solved the green-phase structure single-handedly. But I really wanted this one. I started the project and I identified the black phase as a perovskite. So R. M. Hazen was first author. And if I was first, it seemed obvious that Larry, senior crystallographer at the Lab, had to be second. He had built the equipment, written the computer programs, and had been at the terminal at the moment of discovery. The postdoc Rosses came next, with the promise that they would each be first authors on subsequent, more detailed papers on the nature of the Y-Ba-Cu compounds. Charlie and Dave, whose work was complementary to the principal structural theme of the paper, were listed fifth and sixth. Dave Mao was happy with this order, but asked to have our electron microprobe technician Chris Hadidiacos included as another author. Everyone seemed relatively satisfied with the resulting list: Hazen, Finger, Angel, Ross, Prewitt, Mao, Hadidiacos, Hor, Meng, and Chu.

Then it was on to the abstract. *Physical Review Letters* has a strict seventy-five-word limit for abstracts, a limit that made even cursory description of both black and green phases virtually impossible. But longer abstracts occasionally sneak by. And it seemed essential to give the compositions and salient structural features of the two compounds. So I risked rejection by including details on both phases. The abstract was 140 words long when I was done.

The paper's introduction was problematical in an unusual way. Articles in *PRL* have to be timely and of widespread interest. The first paragraphs are supposed to emphasize this global relevance, even if none really exists. Introductory paragraphs, consequently, are often more exercises in creative science fiction than realistic science fact. In this spirit I wrote an upbeat first sentence.

The description by Wu et al. of superconductivity at 93 K in a mixed-phase Y-Ba-Cu-O compound system is a major breakthrough in materials research.

But in the case of the superconductor no hype was needed. Everyone knew the importance of the discovery and it would be almost impossible to overstate the importance of our research. A mild understatement might have more impact on readers, eager to glean as much information as possible on the superconductor. So the first sentence was revised:

Wu et al. described superconductivity at 93 K in a mixed-phase Y-Ba-Cu-O compound system.

The balance of the introductory paragraph continued in the same restrained tone:

> Their preliminary x-ray powder studies of the multiphase assemblage produced a pattern that differed from other known superconducting structures. In order to understand the origin of the extraordinary properties of this material it is essential to document the atomic structures of the phases that comprise the mixture. We have used single-crystal x-ray diffraction techniques to examine the original material described by Wu et al. (batch 1), as well as additional Y-Ba-Cu-O superconducting samples synthesized at the University of Houston (batch 2). The latter sample exhibits 100% superconductivity by ac susceptibility measurements. In this report we identify the two phases that comprise at least 95% of the sample volume and describe their novel crystal structures.

My principal job was the description of the black-phase structure. We were reasonably confident that this structure, unlike the green phase, was a new structure type. Thus we had to say what we knew about the superconductor clearly and concisely, while qualifying those aspects of the problem that were unresolved—the number of oxygens, the positions of the oxygens, and the possible distortion of the tetragonal structure. This description accounted for nearly half of the text, but it was not difficult to write. A logical sequence of paragraphs quantified the size and shape of the crystals studied, their measured unit-cell dimensions, the details of collecting x-ray intensity data, and the process for solving the structure. Much of that prose was crystallographic boiler plate, only slightly altered for the more general audience of *Physical Review Letters*.

Special care had to be taken in acknowledging the earlier work of Er-Rakho and his French coworkers on the black-phase structure. Not only had they failed to discover its extraordinary electrical properties, but also their structure analysis and description, though close, was wrong. After several abortive attempts I settled on the following:

This structure appears to be related to that of $La_3Ba_3Cu_6O_{14}$ which was reported by Er-Rakho, Michel, Provost and Raveau to have square-planar copper and eightfold-coordinated Ba and trivalent cations on the basis of powder data. . . . Reexamination of their intensity data, however, indicates that the actual structure of $La_3Ba_3Cu_6O_{14}$ is probably the same as that of the superconducting opaque phase described here.

The last substantive paragraph recounted a few of the things that were still unresolved, especially in the details of the missing oxygens. Papers in *Physical Review Letters* are too short to warrant lengthy conclusions, but I did tack on the obligatory proviso: "Additional single-crystal studies are now in progress to resolve some of these structural subtleties."

Acknowledgments and references were all that were needed. We thanked our funding agencies, the National Science Foundation, and the Carnegie Institution in particular, and we thanked Lab colleagues Rus Hemley and Andy Jephcoat for the manuscript reviews that we assumed they'd be happy to do.

The manuscript with figures, tables, and references, all in proper *Physical Review Letters* format, was rapidly assembled and packaged. In record time, at least for earth scientists, we had the paper in the overnight delivery queue, ready for a March 10 "manuscript received" date. Would we be first? We all had visions of the paper being just a day or two late, rejected because it was second. We all knew the heartbreaking stories of scientists who, after laboring for months on a problem, submit a manuscript only to find that someone else has already solved it. And no matter how difficult or elegant the solution, there are no prizes in science for the second person to discover anything. It may not be fair—the intellectual challenge is the same, the creativity and sweat required no less. But knowledge is absolute, and only one person can be the first to know.

That afternoon Art Freeman left another message. Did he really expect me to give him more data? Didn't he realize what he had done—and that Paul Chu knew?

Tuesday, 10 March 1987

I was sure that the worst was over. With two structures solved, we could sit back and relax and get caught up on old business. But I was wrong.

I had forgotten all about setting up our talk at the American Physical Society meeting. The APS session on high-temperature superconductivity was originally Chu's idea. He had proposed a symposium with six speakers: representatives from Zurich, Tokyo, Beijing, and Bell Laboratories were invited to speak along with Art Freeman and Paul Chu. IBM attempted to modify the format somewhat, offering to pay the expenses of the foreign speakers in exchange for an IBM session chairman. But these modest plans began to unravel as soon as news of the yttrium-bearing superconductor discovery became more widely known. Professor Brian Maple of the University of California at San Diego was appointed as a "neutral" chairman, and the format was expanded considerably.

It was only 10:00 in the morning, but in spite of the time difference I thought I'd give Brian Maple's San Diego office a try. He was there, working on superconductors. It was obvious that a lot of people were losing sleep because of elements like yttrium. His manner was firm yet relaxed. He gave the impression of a kindly, diplomatic gentleman. His diplomacy would be necessary to moderate the superconductor session.

Figuring that the University of Houston carried more weight than the Geophysical Laboratory in superconductor circles, I introduced myself as a collaborator of Paul Chu. I told Maple about our work on Paul's original samples and emphasized the unique aspects of the crystal structure. "Can you fit us in?" I asked.

"I see. Well, first I should tell you that the session is already oversubscribed. How much do you know about the meeting?"

"Nothing, really. Paul just told me to call you."

"Well, we couldn't schedule the session during the day because all the rooms are already booked. We have to start Wednesday evening at seven. After opening remarks we have five principal speakers from the five main groups." (Five groups? Which five?) "They each get twelve minutes. Then we have the five theorists, who each get seven minutes—for historical reasons I'm told. Then we have an open-ended session with about forty-five or fifty more papers so far. Each person gets five minutes. I can put you at the end of the list."

I was pleased to get in, but I knew that the only way to have any impact was to be early. Nobody was going to pay attention to fifty or sixty speakers. "I understand the pressure you must be under, but I think it's critical to present the structure work early. We have been hearing all sorts of rumors about this and that structure. Lots of people are basing their conclusions on the wrong structure. Without the atomic arrangement, how can you begin to talk about the physical properties? Can't you put us any earlier?"

I didn't expect to get anywhere with the argument, but he was surprisingly compliant. After a brief pause he told me to wait while he consulted his notes. "The best I can do is put you twenty-fifth, right after a group of materials engineering talks. If you want to have the structure shown earlier you'll have to have Paul Chu do it during his talk. He's third on the program."

I was delighted. Twenty-fifth was still fairly early in the show. "That's great! Thanks for your help."

"By the way," he added, "at least one other group also claims that they have the structure. They'll also be speaking at the meeting."

"Can you tell me who?" I asked.

"I'm sorry, I'm not free to give out that information."

"Well, if you speak with them could you please tell them about my work. I'd be very interested in conferring before the meeting to avoid any conflicts."

He agreed but didn't promise anything. He gave me a few more logistical details and wished me luck. That was that.

I had succeeded in getting a spot on the program, but my mind was in a turmoil. Another group had solved the structure? Already? Who could it be? It didn't seem possible that anyone else could have synthesized Paul's original mix, isolated the black phase, obtained single crystals, and determined the crystal structure in the ten days since Chu's announcement. Perhaps they just had powder diffraction data and were guessing. Or was it possible that someone else had made the yttrium discovery independent of Chu? The uncertainty was maddening.

I didn't have much time to brood, however, because almost as soon as I finished the California conversation the phone rang. While I was talking with Maple a call had come in from Ray Bowers, the Carnegie Institution's editor and sometime press agent. Our downtown headquarters on P Street had heard about the superconductor discovery and they wanted a press release pronto. I called Ray and explained Paul's strictures to him. We agreed to write two texts, one to be released as soon as possible saying the structure was solved, the other for March 18 with the actual structural details. We also agreed to divide the chores, with Bowers preparing the first press release and me providing most of the text for the second. I figured that would be no problem so I set at once to drafting the two-page document.

My writing chores were interrupted by an incoming call on my desk phone. Only a few people had that number. I hoped it was Margee. It wasn't.

I didn't remember having given Freeman that number, but he had reached me. He was calling from downtown Washington at the National Academy of Sciences. He wanted to meet.

"I'm sorry, Art. We're right in the middle of writing up the structure paper. I'm completely tied up."

"You mean you've solved it?" The intensity of his voice raised ever so slightly.

"Sunday afternoon. We got it. The structure is really fascinating." I was going to play this out for all it was worth.

"Can you tell me about it, Bob? You know Cava at Bell Labs is saying it's a cubic perovskite structure. It was in the *New York Times*." By now I understood Freeman's tactic, but I wasn't going to play along.

"Well, that's fascinating. I'll be interested to see if they actually publish something like that. I'm afraid the *New York Times* is not the most reliable peer-review journal."

If Art was getting frustrated he hid it well. "So are they right?"

"Art, all I can tell you at this point is that if they say it's a cubic perovskite they are completely wrong."

"Can you give me details?" he probed.

"Paul has asked me not to tell anybody, and we sent off the paper to *PRL* yesterday. It describes two crystal structures that comprise virtually all of the material. The structure that everybody else seems to be calling perovskite is totally new in detail." I tantalized Freeman without revealing anything useful. "It's got wonderful features that are going to send theorists like you scrambling to figure out exactly what's going on."

"Well, I can't do the work without the structure," he noted casually.

"As soon as Paul says it's all right then I'll release it, but we're a little bit hesitant now. Everybody else is spreading rumors—they're talking to the *New York Times*. I refuse to do that. I've had press people calling me repeatedly over the past couple of days. I prefer to go through the peer-review process and let the information come out then."

I paused, but Art said nothing so I went on.

"I have to tell you, Art, that every rumor I've heard is total nonsense. It sounds like work based on the most sloppy, crude powder diffraction experiments, and I guarantee that no one will ever solve the details of this structure based on powder x-ray work. And if Bell Labs wants to go to the *New York Times* and make fools out of themselves that's their problem!" I was really getting worked up. Actually, I was concerned. Was Bell the group that Brian Maple meant when he said someone else had solved the structure? Cava's outstanding research team knew perovskites as well as anyone and they weren't going to make any obvious mistakes. If they really had isolated the perovskite phase then they would soon solve the structure.

"So the structure isn't the simple perovskite?" Each time Art spoke I sensed a little more tension.

"It's really exciting," I responded, going off at a tangent to his direct question. "Once we got the structure it was immediately obvious how to vary composition in order to vary the superconducting effect. I was surprised to see a structure with such an obvious correlation between an unusual structural feature and an unusual electronic feature. That's what's going to come out when the paper comes out.

"Of course it's all still a little bit of alchemy, don't you think? At least until you guys actually figure out what's going on."

My rambling was really getting to him. "How do you expect me to do it without the structure data?" he complained.

"Of course. You'll have it just as soon as Paul says it's all right. Until Paul actually has a chance to see the paper I can't say anything."

"You know, Bob, now that you've sent off the paper lots of people will know. The *PRL* reviewers have been leaking manuscripts; it looks like some of the big East Coast labs see everything. You might be better off releasing the data yourself."

With that comment Art hit a nerve. I knew the rumors about *PRL*. Some researchers had complained of preferential treatment of manuscripts from the powerful Bell Labs, of systematic unexplained delays of manuscripts by other groups, of arbitrary acceptance and rejection policies, and other disturbing allegations—all virtually impossible to prove. And I wouldn't be too surprised if our structure paper *would* be leaked.

Physical Review Letters had just instituted a special, five-reviewer panel for all superconductor papers. Those five anonymous reviewers, all superconductor experts, had a tremendous advantage in seeing the complete suite of papers that were, by early March, pouring in at a rate of dozens per week. Most of us assumed that at least one of the five unknown referees was from Bell Labs (though we eventually learned otherwise). Was it likely, with billions of dollars at stake, that any industrial researcher would keep important new data confidential from his colleagues?

But that was all part of the superconductor game, and I knew it when I sent the manuscript to *PRL*. "Art, I can't control that. All I can control is what I tell people myself. If Paul is willing to give you the information he has my blessing." I made it clear that our conversation was over.

Freeman was licked, but I didn't feel particularly good about it. For the first time in my professional career I had willfully withheld data from a colleague in need. It was a lousy thing to do. Even if I was adhering to Paul's request, I felt that I was abandoning my own principles. I promised Art that I'd give him a preprint at the New York meeting, if not before, but his good-bye was hollow and depressed.

There was no time to dwell on the implications of Freeman's call. The secretary called again, this time with two messages. A reporter from *Science News* left a number for me to contact, and Paul Chu would call back soon. I decided to leave the line open for Paul's call, which came within ten minutes.

He sounded tired. I could easily imagine how hectic his life was becoming if my own phone messages were starting to back up.

"Hello, Bob. I just wanted to check with you before I go away. I got the paper and it looks fine. I just have a couple of minor changes which I'll send you. I won't be taking any calls after today."

"Things are really getting crazy, aren't they?" I sympathized.

"Yes, it's crazy all right. You know, everybody wants a story. We have reporters everywhere."

"So you're going to get away before the APS meeting?" I asked.

"Yes, until next Tuesday. Then I'll be in New York. I'll need to get a viewgraph of the structure from you. Can you send it overnight?"

"Why don't I just give it to you there? I got into the program. Maple put me twenty-fifth."

"That's good, I'm glad you'll be there. Oh, has the guy from *Physics Today* called you?"

"Not yet, but I've been on the phone so much that I've been missing a lot of calls. I've got to return a call to *Science News*. Should we still keep the structure confidential?"

"Yes, Bob. If you could wait until next week I'd really appreciate it. We've started trying some of those new compositions that you suggested."

"I sure hope something works. I forgot to mention silver yesterday. That might substitute for copper to some extent."

"OK, we can try that," Paul said, pausing long enough to make a note of it.

"By the way," I added, "I just spoke with Art Freeman again. He really wants the structure data. He says we should give it to him because the *PRL* reviewers will leak it anyway. What do you think?"

"Let's just wait until next week, OK? I think that will be best."

"Fine, Paul. I'll do it. You take it easy, and good luck before New York. That's going to be quite a show."

The entire day went like that, with phone call after phone call. Reporters from half a dozen science periodicals called to get the latest scoop. With all the fuss reporters would have to work as long and as hard as the scientists to keep the public informed on the amazing daily developments. I told each caller about what we had done and why, but the details, I explained, would not be released until the big meeting on March 18.

Amid the flurry of superconductor calls one conversation was an unexpected pleasure. Daniel Goodwin, our personable editor at the Smithsonian Institution Press, was calling with good news for Margee and me. The Press had just finished producing our latest book, an illustrated historical study of American brass bands. We had submitted another proposal to the Press in November for a similar illustrated history of fire use and technology in early America, but we had gotten no word. Now they were ready to offer us a contract. That news, coupled with the superconductor success, called for a celebration. I dialed home and made plans for dinner out—just the two of us. I planned to surprise Margee with the book news.

We sat in a quiet corner of our favorite restaurant and talked about superconductors and writing books. Toward the end of the meal Margee surprised me with a new book idea.

"I know we planned to work on this new history project," she began, "but I think you should do something else first. Write up the superconductor story. That's history and you're in the middle of it. The average person would be fascinated."

I was startled by her idea. I'd had enough trouble doing the experiment. To retell it for the public seemed a formidable task. But, if she could hand out assignments, so could I.

"I wouldn't start something like that by myself," I told her. "You'd have to help."

"Me?" She sounded shocked. "I want to *read* the book, not write it."

"Look," I said, "why not come to New York with me next week and observe the physicists in action? If the meeting is as fascinating and historic as I think it will be, we can share our impressions. You can take notes while I sweat out my talk. Maybe we really *could* write something up. We'll stay in the Hilton and explore New York in our spare time."

I paused, waiting for her to agree, but all she would say was, "We'll see."

Chapter 10.

PR

At last we had the information. What were we going to do with it? In saner times and with a more mundane substance the answer would have been clear. Indeed, there would have been no need to ask the question at all for in science, as in other endeavors, there are standard procedures to follow. Write the paper, send the preprint to an appropriate journal and a couple of friends who are experts in the subject, and wait a few months for reviews.

But these were not sane times and the material was anything but mundane. The superconductor was the most exciting new material in decades. Knowledge of the structure could be used in any of a number of ways. So, bearing Paul Chu's restrictions in mind, the question had to be asked: with whom should we communicate and how? Plenty of people had suggestions for us and, as a result, the week before the big New York meeting was almost more exhausting than the weeks of actual experimentation had been. We seemed suddenly to be involved in an impossible balancing act in which responsibilities to fellow scientists, obligations to the Carnegie Institution, and promises to Paul Chu were all at odds.

Wednesday, 11 March 1987

Ray Bowers arrived punctually at 9:00 A.M., as several members of the superconductor team gathered in Charlie Prewitt's office to discuss the Carnegie Institution's press release. For more than an hour we hashed out this phrase and that, saying enough to demonstrate that the structure had been solved without revealing too much. Eventually we all agreed on a three-page version with the heading "Carnegie Crystallographers Contribute to Quest for Superconductor."

The key statements occurred on page two:

> Of special interest was the apparently new structure of the opaque phase, which, they found, is derived from the simple perovskite structure, but differs in several important respects. . . . [It] exhibited unique structural properties that should help theorists to understand the mechanisms of the superconducting transition.

We also listed a number of rumored structures that were wrong, but no other details were given. Those details would have to wait until the second release of March 18.

While we were trying to make news, Larry was busy doing more science. Ever since Nancy's discovery of the French black-phase structure study he had been bothered by many similarities between our model and theirs. Was our solution really better than theirs? Fortunately, Er-Rakho and friends had reported their x-ray powder data in the paper. Thus Larry was able to compute powder patterns based on the two different structure models, and then compare these calculated patterns with the actual x-ray intensities measured in France four years before. By late morning he had the answer. The Carnegie and Caen models were very similar, but our oxygen defect solution clearly matched the French powder x-ray data better than their own. We all marveled at how close Er-Rakho had come, without benefit of single crystals. The implications of the French work soon dawned on us; the lanthanum version of the black phase indicated that the 1-2-3 perovskite structure was not unique to $YB_2Cu_3O_{6+x}$. There were undoubtedly many other compositions with the same atomic arrangement—and any of them could be superconductors, too.

Shortly after lunch I headed downtown to the Washington headquarters of the American Geophysical Union, a professional society that represents a wide range of earth, planetary, and space researchers. I had promised to help in coordinating a few mineral physics sessions, which were scheduled as part of the mammoth annual May meeting in Baltimore. The efficient AGU staff had done virtually everything already; the speaker order, room assignments, and moderators had already been decided. They even managed to avoid serious conflicts between related sessions. All I had to do was sit and listen to the proposed plans.

The discussion seemed routine, with the exception of one item on press conferences. The AGU officials were seeking ideas for newsworthy topics related to Union members. Superconductors, I thought.

For years dozens of mineral physicists had been working on high-pressure silicate perovskites, which are believed to make up more than half of the earth's volume. Many of those earth scientists, including researchers at Berkeley, Princeton, and the University of Illinois, had, at least for the time being, suddenly shifted their emphasis to the new superconductor. It was a classic example of basic research feeding applied research. I made a note to advise the AGU press office about a possible joint news conference in May. I also gave a copy of our structure drawing to an editor of *EOS*, AGU's weekly news and information publication. (The March 24 issue contained the first published illustration of 1-2-3.)

I did not return to the Lab until almost 5:00 P.M., but there was one more task to complete. I thought of another type of chemical substitution that Paul Chu should try. My suggestions so far had been single substitutions: nickel for copper, or strontium for barium. Each of those changes involved two elements of the same electronic charge and similar atom size. I had not thought about coupled substitutions.

In many structures it is possible to replace pairs of atoms simultaneously, providing the atom sizes are similar and the *sum* of the charges is the same. In the 1-2-3 phase, for example, the element pair barium plus yttrium (denoted $Ba^{2+} + Y^{3+}$) might be replaced in part by the atom pair potassium plus zirconium ($K^+ + Zr^{4+}$). I alerted Chu to this possibility as well as a number of other coupled substitutions. I had no idea if the black-phase structure could accommodate even a fractional substitution of elements like potassium and zirconium, but if it worked the superconducting temperature might possibly go up. In any event it was a relatively easy experiment and worth a try.

Besides, letters were now the only way I could contact Paul, and I wanted to keep in touch. I completed the letter before the 6:00 P.M. deadline for overnight delivery and headed home.

Thursday, 12 March 1987

Charlie Prewitt entered my office with a bemused expression.

"Ebert just called. Some of the Carnegie trustees want us to file a patent on the new structure."

James D. Ebert, president of the Carnegie Institution, believed in chain of command. He would never call me directly to ask about this work. Charlie Prewitt, the director, was his middleman. Trained as a biochemist, Ebert was less knowledgeable when it came to the work of earth scientists. Perhaps he and the trustees saw in our superconductor work an opportunity to profit from the physical sciences in the same way his biological colleagues had benefited from gene-splicing technologies. But they evidently didn't understand the proprietary nature of our collaborative superconductor studies.

"Charlie, you know we can't do that." I was exasperated, not at him, but at an administration that seemed so oblivious to the tenor and motivations of our research.

"All right, what should I tell them?"

I answered with intensity. "First, you can't patent a structure type independent of composition. In this case there's no certainty that this structure type—independent of the barium, yttrium, and copper—leads to superconductivity. The copper is probably essential. Second, it would be completely unethical to betray Paul Chu by filing a patent based on *his* material. The answer is no. Absolutely not."

In spite of my vehemence I really understood the administration's point of view. Money was tight and the Institution increasingly relied on government grants and individual donations. Plans for construction of a new telescope in Chile and the new earth science lab in DC meant further drains on limited resources. And in the back of our minds was the incredible story of America's glass industry.

Prior to World War I virtually all high-quality optical glass was imported from Europe, mostly from Germany. Optical glass was a commodity of strategic importance, for many weapons systems relied on telescopes, binoculars, and other such devices. In the early 1910s the total annual American production of optical glass was only a few thousand pounds. With the declaration of war in 1917 the supply of glass to the United States ceased abruptly, and an all-out domestic effort was launched to develop the domestic industry. The government turned to the Geophysical Laboratory for help.

In the first decades of the twentieth century much of the Lab's research focused on the formation of rocks and minerals derived from the cooling of silicate melts—the same mechanism that forms almost all igneous rocks in nature. The tedious procedure began by mixing oxides—SiO_2 plus one or two others—and heating that mixture to melting, usually about 1500° C. Carnegie researchers cooled the mixture slowly until the first crystals, usually of only one mineral, appeared. It was not easy to predict which mineral would form first. The scientists could give the liquid any starting composition by mixing any ratio of oxides, but minerals have fixed compositions. The first mineral to crystallize almost always had a composition *different* from the liquid in which it formed.

Geophysical Lab researchers were not searching for specific practical applications. Rather, they were trying to understand nature. Deep within the earth's crust crystals often separate from the molten rock in which they form; the crystals float away or sink or are left behind as the fluid filters through its porous surroundings. This separation of crystals from silicate liquids of a differing composition accounts for some of the wide variations in natural rock chemistry. That's why Lab scientists studied silicate melts.

But a number of experiments on mixtures of sodium, aluminum, silicon, and other oxides just wouldn't work. No crystals would form. They tried every trick in the book: a small amount of water, different cooling rates, even seed crystals. But the liquid always turned to glass on cooling. As attempts to identify the first mineral to form, the experiments were a dud. But in the process several staff members became experts at how to make glass; those experts turned their attention to the nation's crisis. During the war years more than twenty Geophysical Lab scientists helped companies such as Bausch and Lomb, Keuffel and Esser, and Pittsburgh Plate Glass. By the end of 1918 the annual American production had increased to almost a million pounds per year. It is estimated that the Carnegie Institution spent more than $200,000 on this war effort, but no patents were filed and no recompense was sought. The industrial glassmakers were the ultimate beneficiaries.

Geophysical Lab personnel occasionally recall this story and its lessons. If only the Geophysical Lab had patented glassmaking techniques, if only we had patented Pyrex-type boron-bearing glasses, if only the Carnegie Institution had retained a small fraction of the rights—then our present financial situation would be much easier. But the Institution didn't apply for patents in those days and, as a result, a small army of businessmen and stockholders reaped the financial rewards stemming from optical glass, as well as from hybrid varieties of corn, superhard steels, and other Carnegie advances.

But the superconductor was different. We were the guardians of Paul's samples, and there was no way we were going to betray him. Charlie nodded, and with a simple "I understand" headed back upstairs. Now he had to think of the best way to break the news to *his* boss.

Shortly after Charlie left, the phone rang. It was Margaret MacVicar, part-time vice president of Carnegie and part-time dean and member of the materials engineering faculty at MIT. I immediately assumed that she was calling on Ebert's behalf.

"I understand that you have some exciting news on the new superconductor." Her carefully phrased statement and subsequent pause made it clear she wanted more information.

"Yes, we do. We got Paul Chu's sample a couple of weeks ago and we've solved the crystal structure. It's a new structure type." I was about to tell her more, but I hesitated.

"Can you give me some details?" she prompted. It was a reasonable request from a Carnegie official, but the phone connection sounded long-distance. I had to be sure.

"In what connection are you calling? Did Dr. Ebert talk to you about patents?"

"No, I haven't spoken with him," she replied. "I'm at MIT

with my materials engineering group. We're working on the superconductor, too, trying to devise thin films."

So that was it. Suddenly Margaret MacVicar was transformed in my mind from senior official to just another scientist trying to get a jump on rivals. Our structure data could indeed have helped her in devising methods to fabricate ultrathin layers of the new superconductor. Such "thin films" are essential components of superconducting computer hardware, ultrafast electronic switches, sensitive infrared detectors, and other devices. If Chu's superconducting perovskite could be made into thin films that operated above liquid-nitrogen temperature, it might cause a revolution in the electronics industry.

MacVicar lacked two crucial pieces of information: she had to know the 1-2-3 composition and she had to have the exact unit-cell dimensions. We could have given her the data, but it seemed likely that Paul Chu's group would also be working on the thin-film problem. His demand for secrecy had to be respected. The information was not yet public.

"I'm really sorry, Margaret, but Paul Chu made us promise to say nothing to anyone about the structure until after the March eighteenth APS meeting. The best I can do is send you a preprint to arrive on the nineteenth." I tried to sound as apologetic as possible, without leaving any doubt about my resolve.

"I understand, Bob. I would appreciate the paper as soon as you can send it. Good luck with your work." She seemed cordial enough in her good-byes, and I could only hope that I hadn't offended her.

It had been a long morning and I felt tired and hungry. Looking at my watch I realized why. It was 12:27—definitely time for lunch. Unlike Washington's other Carnegie laboratory, where the lunch hour is spent in a communal meal, the Geophysical Lab researchers disperse at midday—to local cafes, to jogging trails in the park, to the volleyball court. The brown baggers usually meet informally in the Lab's seminar room. That's where I headed, and I was pleased to see Larry, Rus, Andy, Charlie, and the Rosses already there. It was the first time in almost a month that so many members of the mineral physics team had eaten together. The large table was almost covered with plastic wrappings and the usual assortment of yogurt, cheese, and sandwiches.

"So who's going to New York next week?" I asked as I sat down and took a bite of my standard turkey-and-pickle-relish sandwich. The annual APS meeting had been of interest long before the superconductor business because of a special session on high-pressure research on gases. Several of us were speaking.

"Dave, Andy, and I are scheduled to talk at that Thursday morning session," said Rus. "And you're talking on hydrogen, right, Bob?"

I nodded yes while eating my sandwich.

"Are you going, Larry?" Ross asked.

"No, I'll be in Austin for the American Crystallographic Association meeting—talking about hydrogen. I guess I'll miss all the fun. So what are the plans for the superconductor extravaganza?"

"It's amazing," I said. "I talked to Brian Maple, the organizer, last Tuesday. He said there were already about sixty talks, with more added every day. It's going to run all night. He's got us scheduled twenty-fifth."

Rus whistled. "That's bizarre. I've heard of all-nighters to get work done. Hey, I've done enough of them myself in the past couple of weeks. But I've never heard of an all-nighter to *present* the data."

"That's not the only strange thing," Larry added. "We got the samples when?—the last week of February?—we write the paper about two weeks later, and present the paper in less than a month. It's unheard of. I just hope we haven't gone too fast and made a dumb mistake."

"It's different, all right," I agreed. "Has there ever been anything else like it, with so many people working so hard on one material?"

Charlie had been quiet, but he answered right away. "The first moon rocks were sort of like this. Everyone got samples at about the same time and for several months it was crazy. People lived in their labs and worked twenty hours a day. But there really wasn't any money at stake. Here we're talking about big business as well as science, so all the engineers are working like mad at the same time as the scientists."

"Hey, that reminds me," I said. "A guy from *Science News*

called. In the course of the conversation he wanted to know what kind of scientists worked at the Geophysical Lab."

"Tired ones," Larry offered.

"Yeah, tell me about it." I laughed. "But he meant like a geologist or geophysicist. What do you call yourself?"

"A crystallographer," Larry replied without a second thought.

"A chemist," said Rus just as fast.

We were working our way around the table so Charlie spoke next. "I usually say materials scientist or an earth scientist. It really depends on who's asking."

The postdocs had been pretty quiet during these soul searchings, but Andy Jephcoat was ready for us. "I've got no problem there; I just say I'm single."

The afternoon was relatively uneventful, with the exception of a memorable phone call. I suppose I shouldn't have been surprised that it was Art Freeman again. This time he was anything but calm.

"Bob! Thank God I've reached you. I have to know your latest structure data. We're running out of time. Please, help me."

I was silent for perhaps half a dozen seconds, unsure what to say. There was no point in beating around the bush—I had no choice but to confront him. "I don't know what to do, Art," I replied. "Paul says that you've filed your own superconductor patent, and that you've given our preliminary structure results to the guys at Argonne. Is that true?"

Art didn't respond at once; when he did speak his voice sounded tired and resigned. "Damn superconductors! It's all a terrible misunderstanding. There's no patent. I would never do that to Paul. And I didn't give any data to Argonne. All I did was tell them about the Chinese newspaper that had the yttrium information. In fact, they're angry at me now because I *wouldn't* tell them any of Paul's secrets. And now you won't talk to me either. Someone's been lying to Paul."

I'm not sure why, but I knew he was telling the truth. I thought back over my dealings with Freeman and it was clear that he had always been intensely dedicated to his science. There was never any hint of deceit. He was one of the good guys and I regretted having mistrusted him.

I couldn't blame Paul Chu. He had been under tremendous pressure to reveal his secrets. Scientists from industry and academia had besieged him with inquiries, while government officials probed him for data. He had evidence of an industrial spy within his own Houston physics department, not to mention the suspected leaks to Beijing. Paul knew the importance of our findings and, given the rumors of Freeman's Argonne connection, it was perhaps prudent to block Art's access to our structure results.

But the data were in my hands and I knew Art Freeman was telling the truth. Then and there I told him everything— all the details of the superconductor structure that he needed to perform his calculations. It was a decision I did not regret.

I could not have blamed Art for being upset or resentful at my mistrust, but he never uttered a word of complaint. Instead he focused on the scientific implications of the structure work. He immediately grasped the importance of the copper-oxygen layering but he didn't waste time speculating over the phone. He had work to do and time was running out.

Friday, 13 March 1987

The last in-house seminar before New York was supposed to be a command performance with our superconductor structure making its debut at 4:30 P.M. But when Carnegie officials and scientists from outside the Institution heard rumors of the talk and asked to attend, we were forced to reconsider our plans. We had promised Paul that we would avoid any public announcements of the structure until March 18 and there was no way to make good on that promise if our presentation attracted a large contingent from Washington's scientific community. Reluctantly we decided to postpone the lecture for a week and to gather instead for an informal discussion.

The crystallographic team was relaxed and self-satisfied as we lounged in the seminar room that afternoon and drank our beer. We were also, I'm afraid, a little insensitive to the feelings of our coworkers who were suffering the double indignity of seeing their own work overshadowed by superconductor media hype, only to have their curiosity about our work snubbed. As I spoke casually with various colleagues I half expected to be taken to task for deliberately withholding the details of a Lab experiment from fellow workers. What I didn't expect was outright hostility toward The Experiment itself.

Norwegian-born geochemist Bjørn Mysen voiced the resentment that several of our colleagues obviously shared. He began with his usual debating tactic, a hard-nosed rhetorical question: "Given the fact that this superconductor is not an earth material, are we *really* justified in spending time on it?" Before allowing us to respond he drove the barb deeper. "After all, if you want to publish in the *New York Times* shouldn't you be in another field?"

We were all a bit taken aback while Bjørn, obviously enjoying the reaction, leaned his tall, lanky frame back in his chair and lit his pipe. Though only forty, Bjørn had made more than his share of scientific enemies with his strong, sometimes controversial opinions and fierce confrontational style. But he was well respected at the Lab for his keen intellect, prodigious output, uncompromising ethical standards, and delightful Scandinavian accent. His point deserved rebuttal, though none of us was eager to take him on.

It was, surprisingly, Rus Hemley who picked up the gauntlet.

"Come on, Bjørn," he countered, "this is one of the biggest breakthroughs in materials science in years, and it just fell into our laps. What should we do, say no? Besides, the stuff is perovskite-related and that's Geophysical Lab turf."

Mysen wasn't satisfied. "You certainly didn't know about the perovskite part when you started. All you knew was that it was going to make somebody rich and famous." Bjørn's underlying point was becoming clearer and Ross Angel didn't like it.

"I never thought the Geophysical Lab was supposed to be an ivory tower. Are you saying we shouldn't work on problems that have practical applications or are interesting to the general public?" Unlike Rus Hemley, Ross was defensive and angry.

Bjørn knew he had struck a nerve. "So I suppose you're going to drop everything else for the next couple of years to work on this? I think that would be most ill advised."

I glanced around the room; most people were enjoying the animated exchange. Charlie Prewitt, who would never get involved in such a debate, was smiling with arms folded across his chest. Larry laughed, nodded, and raised his eyes heavenward as Tom Hoering whispered something in his ear. Tom, I was sure, was too much of a pragmatist to accept Bjørn's point of view. Others were listening intently while ignoring their beer.

But I sensed the danger of harsher words between Ross and Bjørn. A new perspective was needed. So I interrupted.

"No, no, Bjørn. We don't have any plans to do anything unless Paul Chu sends us new samples. The big industrial labs have hundreds, maybe thousands of people working on it. They'll be taking over. We just had some fun with it for a month or so. Hell, it was fun—wasn't it?" I added to no one in particular.

That seemed to defuse the situation temporarily and we went back to talking in smaller, more relaxed groups. But I couldn't help wonder how many other scientists in the close-knit earth science fraternity resented our involvement in a field that was rapidly becoming a media event.

Saturday, 14 March 1987

I was in a restless mood, spending a couple of hours preparing viewgraphs to illustrate our results in the clearest possible way for the upcoming meeting. We had checked and rechecked our numbers amid the daily rumors of other groups with other solutions, some radically different from our own. Were we wrong? Could we have missed something obvious? The uncertainty haunted us, because while history merely ignores the also-rans, it can be merciless to the bunglers.

When my desk phone rang I welcomed the diversion from these dark thoughts and did not hesitate to answer. It was Margee and her voice cheered me immediately. So did her news. She had arranged her schedule and had decided to come to New York with me.

Sunday, 15 March 1987

Dave Mao had heard again from Paul Chu. His Houston group had made a new superconductor with a much higher critical temperature. If they succeeded in reproducing the results they promised to send their new sample to us immediately. Most of us were hanging around the Lab anyway, more from habit now than any urgent work. But as exciting as the prospect of new temperature records might have been, we could only groan. Were we really expected to solve this new structure in the two days before the APS meeting?

Monday, 16 March 1987

We spent the day waiting for a package that never arrived. We weren't about to start any new projects on the three idle diffractometers. We had to be ready to move fast. But Paul's call must have been a false alarm.

Lunchtime conversation in the seminar room shifted gradually from speculation on the new samples to an intriguing conversation on superconductors and society. None of us was very adept at investments, but it was fun to muse on superconductors and their potential impact on the economy. A few specific activities, we agreed, would thrive—yttrium mining and liquid-nitrogen manufacture, in particular. But no company made more than a tiny fraction of its income on those commodities. There were no good investments there.

The most likely economic effect it seemed would be a long-term shift in energy production from costly, localized electrical generators to remote giant power plants. Desert solar collectors covering hundreds of square miles, remote nuclear power plants, and hydroelectric or geothermal facilities far from population centers were possible consequences of superconducting power cables.

"So," Charlie mused, "should we rush out and sell oil stocks short?"

Ross had obviously thought seriously about that: "Not right away; it will probably be years before anything major happens."

"I guess that means we're not all going to get rich," Nancy Ross joked.

"Well," Ross quipped, "we'll just have to settle for famous."

Without any immediate experimental goals, and without knowing the status of the new samples, the day passed slowly.

The viewgraphs for New York were ready and I had carefully prepared my five-minute presentation to cram as much data as possible into the impossibly short talk. I left the Lab at 5:00 P.M. with the feeling that nothing in my life would be settled until after that Wednesday night affair.

That evening Margee and I began to pack our bags for New York. I couldn't resist adding to the clothes pile a favorite bright kelly-green shirt with the basic perovskite atomic structure boldly depicted in white. For a moment I even thought of wearing it to the all-night session, but I quickly decided against making that kind of lasting impression on the physics community.

As I moved about the room Margee could sense that I was agitated about the upcoming presentation.

"Why are you so worried? You solved the structure, didn't you?"

"Solved? It's not a simple question of right or wrong. It's more like right or more right."

We had found a structure model that explained our data reasonably well, so we were right in that sense. But with better crystals or more time we'd have obtained better data and a better model. And no matter how well we did, someone else was going to do it better next month, and then better still next year, and so on and so forth. So what was right?

The bags were as packed as they were going to get that night, so I picked up a sci-fi novel as a distraction. I was still reading in bed at 9:30 that evening when the phone rang. It was Larry, calling from the American Crystallographic Association meeting at the University of Texas in Austin.

"Hi, Bob!" he said. "I thought I'd better warn you. I spilled the beans." Then he told me the story.

Finger had taken a viewgraph of the superconductor structure to the meeting, where he planned to show it on Wednesday afternoon as a sort of sideshow to his main presentation on hydrogen at high pressure. That way, we figured, the crystallographers in Texas would get the news at the same time as the physicists in New York. But things didn't work out that way at all.

Hugo Steinfink, distinguished crystallographer at the University of Texas, had also worked on the new Y-Ba-Cu-O superconductor. As one of the organizers of the Austin meeting, Hugo was able—at the last minute—to set up and announce his own special talk on the new structure. The informal presentation began at 4:00 P.M. in a small university lecture room that was crammed to capacity. It appeared to Larry that virtually every one of the two hundred inorganic crystallographers at the ACA was present: Sam La Placa from IBM, Sidney Abrahams from Bell Labs, Jack Williams and Mark Beno from Argonne—all of the big boys were there. All of our superconductor competitors were there.

In his fifteen-minute, unillustrated discourse Steinfink explained the procedures by which his group had determined the composition of the black phase and had deduced many of the structural properties. Like us, they had found a tetragonal unit cell based on a tripled perovskite repeat unit. They, too, had discounted the Er-Rakho structure model. They, too, had observed a regular Ba-Y-Ba ordering pattern. But Steinfink's powder data and preliminary single-crystal work would take him no farther; he was unable to say anything about the oxygen positions.

The discussion session was a wide-open affair, with numerous questions and few answers. Without a clear illustration of the structure, it was difficult for the audience to make much sense of which atom was which in any of the several possible structural variants. It was too tempting for Larry, who carried the Geophysical Lab viewgraph in his briefcase. Without fanfare he unveiled the structure drawing, more as an aid to Hugo than as an exhibit of our work.

It took the audience only a moment to absorb the implications of the illustration. The fabulous copper-oxygen planes were boldly highlighted in the colored diagram. Larry was no showman, and he certainly wasn't about to upstage his good friend Hugo, but the focus immediately shifted to the projected image.

"How did you determine the space group?" "What is the total oxygen occupancy?" "Why is the yttrium-level oxygen missing?" The questions came furiously as the assembled masses soaked up details of the unique structure.

What could I say to Larry? I am certain I would have done exactly the same thing—possibly in a way less considerate of the speaker. So I just said, "That's fine, Larry. It sounds like it was a fascinating discussion. I wish I had been there."

But my actual thought was that we were in a real fix. Everyone—at least everyone who mattered—now knew our results. And we knew nothing of theirs. In crystallography, as with almost any other scientific discipline, there are seldom answers that are absolutely right. A structure model is, at best, an approximation: the best match the crystallographer can find between observed x-ray patterns and the patterns calculated from the structure model. The match is never perfect, and there is always a chance that a better, radically different solution exists. Just as we were able to invalidate Er-Rakho's solution, so might our hard work be discounted.

In one sense we knew that our work would soon be superseded. Groups at Oak Ridge in Tennessee, Naval Research in Washington, Argonne in Illinois, and other government facilities were already starting to collect neutron diffraction data. Oxygen is a strong scatterer of neutrons, so neutron diffraction, though time-consuming, was obviously the way to go. In another few weeks those groups would be able to report oxygen positions and occupancies with much greater precision than any x-ray study. We knew, in effect, that our work would be obsolete before it was published.

But the New York meeting was another matter. That was our chance to contribute something original and new. For one brief, glorious moment we would share the knowledge that we had fought so hard to win. I could only hope that we hadn't given away too much, too soon.

Part III.

NEW YORK

Chapter 1.

Metropolis

Viewed from a distance, across the brittle brown marsh grass of New Jersey, the city was quiet and still and very beautiful. I'd been to New York often enough as a teenager to know that this perception was largely an illusion, but it was a pleasant one and it suited my introspective mood. Soon enough the train would hurtle through the Hudson River tunnel and inject its passengers into the heart of the city. I welcomed these last few moments of peace to review my talk for the evening's superconductor session.

Margee, reclining comfortably in the seat next to me, grinned as I checked the notes in my briefcase again. I had been over them perhaps a dozen times already during the journey north from Washington, but at this point I was scanning less for detail than for reassurance that all was in order. I decided that it was. Although I would have preferred to use slides to illustrate the structure, I was in the world of physics now and was thus constrained to use viewgraphs, those versatile lecture tools that cleverly allowed a speaker to alter or augment data at the last minute. Not that I intended to make any changes in my presentation. Our lab work was complete. How our data related to the larger superconductor picture was another question altogether, but I would have to wait until evening for the answer to that one. I sighed and replaced the papers in my briefcase only seconds before the train pulled up to a dimly lighted platform in Pennsylvania Station. We had arrived. The big day had begun.

The sounds of New York assaulted us the moment we entered the cavernous Penn Station waiting room. Blaring public announcements, punctuated at regular intervals by the chattering of a pneumatic drill, echoed stridently above the surging noises of travelers on the go. It was only 11:30 on a Wednesday morning but the terminal was crowded. Instinctively I gripped my briefcase tighter and began to jockey my way along, carefully angling the luggage to fit the spaces available. Margee maneuvered into the gaps I left behind me, and in this halting way we progressed slowly down a corridor, up the stairs, and out onto Seventh Avenue.

It was not quite as crowded there, but the raw energy of the city streets hit us like a gale at sea. Speeding taxis, jeering horns, and flashing neon signs all competed for our attention. Even our lunch spot, an express Chinese restaurant decorated with bold black and red stripes, seemed to capture the essence of the city: it was fast, noisy, and geared for the masses. Some physicists later remarked that this churning, stressful environment was a poor locale for the American Physical Society meeting, but I had to disagree. Just being on the city streets helped clear my brain and got the adrenaline pumping. By the time we had walked the twenty blocks north and pushed through the doors of the imposing New York Hilton I was ready for anything that science, on the cutting edge of a new age, could offer.

Good thing, too. The convention headquarters was like Penn Station with an intellectual motif. Small groups of scientists engaged in earnest conversations clogged the lobby, and I thought I overheard the word "superconductor" several times as I made my way to the registration desk. Our room was not ready, but I had two phone messages that needed attention in the meantime. One was an urgent request to contact Ray Bowers, the Carnegie Institution editor who had helped me prepare the press release. The other simply read "Call 1828."

Who could it be? Maybe Paul Chu, since I had to give him the structure viewgraph. We decided that I'd call the number while Margee scouted out the registration area and APS message board in the upstairs lobby.

As she disappeared into the crowd, I went off to find a house phone. I quickly discovered a whole row of them and every one was in use. I waited for what seemed like a long time, but probably was only a couple of minutes. When it was my turn I dialed the enigmatic room number. Busy signal. I waited a minute and tried again. And again. But no luck. Whoever was in room 1828 was obviously in demand. I headed across the corridor to the pay phones and tried Ray Bowers's number. This time I got through.

"Bob, I'm glad you called. Have you handed in the press releases yet?" He sounded worried.

"No," I replied, "I was going to do that after I registered for the meeting. Is something wrong?"

Bowers told me the problem. Charlie Prewitt had found a couple of small errors that had to be corrected before the releases were handed out. Ray detailed two places in the text where we had transposed the ratios of the three metal elements. I grabbed a copy of the three-page report from my briefcase and found the spots in question. Sure enough, they were small errors, but ones that might translate into significant confusion for a journalist. We would have to correct all 200 copies by hand—not my idea of a fun way to spend time in New York City.

I said good-bye to Ray and realized that the promised ten minutes had elapsed so I rushed upstairs to the registration area to find Margee. Her news didn't ease my anxiety. On the bulletin board she'd found another message that read, cryptically:

Pls call me
Rm 1828
Bob Hazen

What was going on? It was the same number as before, but now signed with *my name?* Surely there was no reason for the secret message writer to disguise his or her identity. Perhaps it was written by someone who was distracted and tired.

We went back to the house phone and tried again. This time the call went through. Though the voice on the other end sounded weary, I had no trouble recognizing it as belonging to Art Freeman.

"Bob, thank goodness you called," he said.

"Oh, Art. So you're the mystery man in 1828. This place is a circus. What's up?"

"I've got to see you right away. I've got wonderful results that you must see and we have to talk about Paul. Come up to my room." He sounded agitated and I felt myself responding in kind.

"Is something wrong with Paul?"

"Come upstairs. We'll talk about it."

We made our way along the thickly carpeted corridor to Freeman's room and knocked quietly at the door. He answered immediately and let us into a room that would have looked like a standard Hilton accommodation were it not for the debris scattered everywhere—papers were thrown all over the bed and viewgraphs spread out on the table and chairs. Art seemed preoccupied and tense, but he relaxed long enough to say a warm hello to Margee as I introduced them. He shuffled some papers so she could sit on the edge of the bed. Then he introduced us to his hard-working graduate student, Jaejun ("J-J") Yu, who had been standing by the window reading a report. As the young Korean scientist came over to shake hands I recognized in his exhausted expression the all-too-familiar signs of too many all-nighters. Freeman was quick to explain why.

"J-J has been working round the clock since our conversation about the new structure last Thursday. He left Chicago at 3:00 this morning with the latest results. I've got to show you this."

He hustled over to a viewgraph-laden chair and selected two cheerfully colored diagrams. "Look at this—it's incredible!" Art proceeded to explain his latest band structure calculations which, though largely incomprehensible to me, seemed to confirm that the distinctive layer structure was directly related to the superconductivity. "J-J and my postdoc Sandro Massidda had to work nonstop to finish these."

I could only nod my head and compliment the two physicists on their results. Then it was my turn to show Art what our own lab had done.

"Here's a viewgraph for you. This is the structure." I handed him a version of the crystal drawing with copper-oxygen layers highlighted in bright red. "You're welcome to use this in your talk, but I need it back before mine."

"That's great, Bob. This will be very helpful." Art added

the illustration to his viewgraph piles, which looked far too high for the upcoming short talk.

"It's been quite a time, hasn't it?" I paused and added, "You look plenty tired yourself, Art."

There was a brief lull in the conversation. I suppose we all were thinking back over the extraordinary weeks that had just passed. But our conversation was not yet over.

"So what's going on with Paul?"

Art Freeman sighed and began the story. "I'm really upset about the press releases from other groups; they seem to play down Paul's role. You have to understand, Bob, that there's a lot of resentment against Paul. He's been extremely successful but he doesn't do things in a conventional way. He doesn't work for one of the big laboratories and he's filing patents through his university. Now he's made this yttrium discovery —snatched it right out from under the noses of IBM and Bell Labs.

"There's going to be a Nobel Prize for high-T_c superconductors—maybe more than one—everyone knows that. It seems obvious that Müller and Bednorz will share part of the prize, but the real question is who else will be included. The Japanese? The Chinese? Bell Labs? Whoever gets that prize is going to gain incredible prestige, not to mention an awful lot of credibility when it comes to patent rights. Paul is an obvious choice for the prize, but a lot of people want to prevent it."

"But why should they?" I asked. "How can they argue against him? He was clearly first to break 77 K. And his discovery of the pressure shift, alone, is of Nobel caliber."

"Of course he was first—by several weeks. He blew everyone else away. But now some people resent his keeping the composition secret for so long and others, who worked on ytterbium instead of yttrium, are mad because they wasted time on the wrong stuff. And then lots of people are just plain embarrassed. Think about it. The guy isn't from Harvard or Stanford or Berkeley or Caltech. He's from Houston! The stuff was first synthesized at the University of Alabama, for God's sake. Not exactly your big-name physics schools.

"Look at it from the industrial labs' point of view. IBM could argue that Paul's work was just an obvious follow-up to the original thinking of Bednorz and Müller. Others can claim that Paul just stumbled on the yttrium compound by accident, and that the 90-K superconductivity was more a lucky accident than a legitimate original discovery. Bell Labs can hope to get part of the prize for their physical characterization work and rapid implementation of the materials—and you have to admit, nobody's better than Bell Labs at that. The Japanese and Chinese groups can claim that they made the same yttrium discovery independent of Paul, and who's going to prove they didn't? You know, it's a damn good thing that you guys figured out the identity of the superconducting phase; without that information other groups might be way ahead of Paul by now."

"So what can we do to help Paul at this point?" I asked.

"Just give a good talk tonight and thank Paul for the discovery. I don't know what else to do."

I wasn't sure what to do either. There was little I could do to help Paul in his fight for recognition; time would have to sort out the winners and losers. But I resolved to begin my presentation by giving Paul full credit for the discovery of the superconductor and thanking him for the chance to work on it.

We all had much to do before the evening, so Margee and I said good-bye to Art and J-J Yu and headed back to registration.

Our twenty-fourth-floor room was a relief from the bustling lobby and elevators, but we weren't in a position to relax. The press releases had to be revised and taken to the press room. Margee and I did the job assembly-line fashion, with her correcting the second page and me fixing the third. In our hyped-up state of mind it didn't take all that long to make the changes; by 2:30 that afternoon we were heading down to the sixth-floor press headquarters.

The press room was a shock. Table after table stacked high with releases filled the room. Dozens of reporters milled about, sampling the bewildering array of scientific offerings. Just a brief glance at one table confirmed my suspicions: most were on superconductivity.

As Margee waited just outside the door I tried to find an APS official to hand in my own contribution to the news glut. I was directed to a middle-aged man with an official-looking blue badge. He assured me that he could take the releases and distribute them. First, however, he wanted me to describe the contents.

I enthusiastically told him of our work with Paul Chu, a name I assumed would garner respect. But before I could point out the significance of our work in isolating the superconducting phase and determining the crystal structure he interrupted.

"We've been looking for Paul Chu!" he exclaimed. "CBS television wants to do an interview and no one can find him. Where is he?"

"I haven't seen him today," I responded, somewhat nonplussed at his evident lack of interest in my science. "He said he'd meet me tonight at the meeting."

"Well, if you see him, make sure he comes to the press room." He then paused long enough to look at the title of our release: "Carnegie Crystallographers Announce Atomic Structure of Superconductor."

"So how do your structure results compare with the Bell Labs, Bellcore, and IBM announcements?" he asked.

I was stunned. There were *three* other structures? I should have been prepared, but I could only stammer that I hadn't seen them and that I'd like to get copies of their press releases.

"I'm sorry, but there just aren't enough copies for that. All press releases are strictly for members of the press. In fact, you shouldn't even be in this room without a press badge. Who let you in?"

My mind was in a turmoil as I was politely ushered to the corridor. Three other groups had structures! How was it possible? I had expected to be the only one with data on the unique 1-2-3 perovskite variant. I had hoped to steal the show, as thousands of physicists strained to absorb every detail of *my* structure. Even if there had only been two of us, the Geophysical Lab work would stand out. But four structures! At best, it seemed to me, our hard-earned results would be lost in the shuffle. At worst, I feared, we could be proven wrong.

I had to get away from the hotel. Even the New York streets seemed a release from the intensity of the hotel. Margee and I walked more or less in the direction of Central Park, without any clear destination in mind. After a few blocks we found ourselves in front of a favorite art gallery on 57th Street. We stood on the sidewalk and debated whether to go in, but I knew instinctively that the unhurried pace of the gallery would settle my nerves more than a walk in the busy park.

With soft lighting and soothing gray walls, the gallery provided a quiet sanctuary for an hour or so. Beautiful paintings were all around, but we were strangely drawn to one diminutive, gemlike oil by the Hudson River master, John Casilear. It was a deep woodland scene with shafts of light streaming through the trees. A solitary, red-shirted figure reclined, dwarfed, before a mighty roaring waterfall. He sat motionless, awed at nature's incomprehensible grandeur.

As the minutes passed, the reason for my captivation with the painting gradually dawned on me. I understood that man. All of us who have marveled at nature and who have labored for a shard of knowledge understand that man. From that vantage point the tensions and rivalries of the past month seemed so trivial and needlessly divisive. And I knew that there was no need to fear what lay ahead that evening.

The doors weren't scheduled to open until 6:45 P.M. but as early as 5:30, a full two hours before the session was to start, people began to gather in the corridor leading to the ballroom. Margee and I, having picked up a couple of roast beef sandwiches to eat while waiting in line, joined the group just before 6:00 and took up a position on the fringe of the action. It was a great spot in which to deposit our lunch bags and enjoy the carnival-like atmosphere—and we held onto it for all of two minutes. Scores of additional spectators soon appeared and we found ourselves not at the edge of the crowd but at its pulsing center. Ten minutes after that we were, strictly speaking, closer to the front than the back of the mob, but that was an irrelevant distinction. All that mattered was that there were people packed in around us in every direction, closing in on us like a vast humanoid straitjacket and making it impossible to move an elbow without jabbing a neighbor. Dinner would have to wait.

It didn't really matter. It was a friendly, good-natured crowd. Everyone was keyed up and enthusiastic, energized both by the knowledge that they were participating in a historic event and by the emotional high that comes from sharing something important with others who understand.

The largest Hilton ballroom had already been committed to a wedding party, so the superconductor session would be held in the smaller Rendezvous Trianon Ballroom, which had seating for only about twelve hundred people. With the crowd swelling larger and larger it was obvious that more physicists would have to hear the papers via video monitors in auxiliary seating areas than in the main hall itself. But except for an intense West German physicist (who remarked that his country's constitution would have allowed the physicists to commandeer the larger ballroom from a wedding party on the grounds of national priority), everyone we met seemed resigned to the mob scene and the rowdy stampede that was sure to follow.

To be sure, there was an inherent danger in the densely packed assembly. At one point, a full thirty minutes before the doors were scheduled to open for the public, a technician passed through the portal. Several overanxious members of the crowd pushed forward, sending a moderate ripple of pressure through those of us near the door. Much more distressing was the plight of a half-dozen scientists off to our left. Hoping to join the crowd near the front of the wide corridor, a number of physicists had descended from the floor above on an escalator that disgorged its passengers, one after another, into an area that was already jammed to capacity. Eventually there was no more room, and the down-going party crashers found themselves approaching a solid wall of humanity. One man fell down at the bottom, and the others had to scramble up the down escalator to avoid injury before the alert hotel staff was able to turn off the treacherous device and block off that access to the ballroom level. But other than those few dramatic moments, the waiting was marked only by the subdued din of a hundred conversations. Then at last, the doors swung open.

The multitude let out an enormous cheer and surged forward with a growing force. We were slowly pressed together, as the crowd began to move. Several humorists in the group made cattlelike "mooing" noises as they were unceremoniously herded toward the narrow opening. The worst of the pushing and shoving began as we got close to the doorway. Physicists met the laws of physics head-on, as the mob that had filled the forty-foot-wide corridor had to squeeze through an eight-foot-wide hole. As Margee and I closed in on the gateway our speed picked up. Faster and faster we were shoved and jostled until at the very point of entry we had to run and stumble forward to avoid the trampling masses behind. It was a brief moment of idiocy, near panic, as the mindless crowd swept forward.

At last we made it into the ballroom. Everyone who had waited for so long near the front of the line assumed that having gotten that far there would be seats for all. But we in the vanguard quickly discovered that other crowds—just as large and just as eager—were pouring in from other doorways, all converging on the same small ballroom. It was every scientist for him- or herself as the undignified rush for seats began.

I wasn't too concerned about myself because there was a roped off area for speakers, but I hoped that Margee could find something nearby. I underestimated the competition. By the time we were in the ballroom proper all visible seats were occupied and even the last row of the speakers' section had been appropriated by frustrated physicists who surreptitiously repositioned the red velvet roping. In the end Margee sat on the floor next to my chair near the back and at the far right side of the reserved section, and there she remained, recording history with notebook and tape recorder in hand, for much of the evening.

Not everyone was so lucky. The fire marshal was more than a little concerned at the chaos that had crammed every corner of the hall—seats, aisles, windowsills, and doorways —with highly educated humanity. More than two thousand scientists had taken up temporary residence, with countless more still trying to squeeze their way in. The situation was out of control and obviously unsafe, and the overcrowding would have to be rectified before the meeting could begin.

Shortly after 7:00 P.M. Brian Maple, the session moderator, made his first plea to clear the aisles. No one appeared to budge. Only after repeated requests, with increasingly stern warnings that the entire meeting would have to be canceled without cooperation, did order eventually prevail. Hundreds of reluctant scientists had to retreat to adjacent rooms where video monitors would display the sessions. They joined the rumored thousands of others who never even made it into the main hall.

As the aisles gradually began to clear we relaxed, waited, and surveyed our surroundings. The Trianon Ballroom was a large, elongate rectangle, perhaps four times as long as it was wide. At one end of the hall, facing the audience, was an elevated table with six chairs, six microphones, and an abundance of water pitchers and paper cups. Evidently groups of speakers would sit at that table to await their talks and answer questions.

To the right of the table was the moderator's podium, while the speakers' podium and viewgraph projector were symmetrically disposed to the left. From the angled, corner position of the speakers' stage it was obvious that only a small fraction of those in the main hall would be able to see the talks directly. Therefore, close to the front on either side of the main aisle were two massive video cameras, poised to transmit the proceedings to large screens situated at several spots outside the meeting hall, as well as at regular intervals down the side of the long ballroom, itself. The resultant videotapes, we had been promised, would also form a permanent record of the proceedings. Each of the speakers had to sign a rather formal-looking waiver allowing the distribution of the videotapes by the American Physical Society.

As I scanned the rapidly filling speakers' section I recognized only Art Freeman, who engaged in animated conversations in the third row of chairs. This was my first APS meeting and I didn't know any of the designated speakers except for my recent acquaintances, Art Freeman and Paul Chu. So rather than participate in extended scientific dialogues, Margee and I just sat and reflected on the amazing scene. As we talked we both scribbled notes on the unique setting and the distinctive crowd around us.

Shortly after 7:00 I spotted Paul Chu, who had wisely waited until after the big push. He made his way over to us, shaking hands and exchanging greetings along the way, and sat down next to me. He seemed delighted to meet Margee, and he again complimented the Geophysical Lab for its work. But in spite of his graciousness, Paul was obviously on edge. He kept his left hand tightly clenched on his folder of viewgraphs and his expression was set and firm. Though he hadn't eaten in hours, he declined my offer of half a roast beef sandwich. I figured he was just being polite so I offered a second time. "It's going to be a long night," I told him. "You might need it."

Paul laughed. "It's not a battlefield," he said, adding that he felt too nervous to eat.

But he did accept my crystal structure viewgraph, a duplicate of the one I'd given to Art that afternoon. Paul promised to show the illustration in the course of his talk. Then the three of us simply chatted quietly about the extraordinary month that had just passed.

Though his face appeared hard-set, he talked in his usual friendly, relaxed way. Only when I mentioned the yttrium/ytterbium controversy did he become overtly agitated.

He was incensed at the entire handling of the affair by the *Physical Review Letters* office. "*No* information about *any* paper should be released or tested experimentally prior to its publication," he insisted. "I never even knew the error was there. So later on I got so many calls from people saying they tried ytterbium without success. So I said, 'My God, why should they try ytterbium?' I went back and looked at the manuscript and, oh boy, there it was. So I changed it immediately. It was all a simple mistake."

"But no one wants to accept that it's an accident," I sympathized.

"It *was* an accident. Even if many Bell Labs people say otherwise. You tell me, how is it possible that so many Bell Labs people *knew* that Yb was in the original manuscript if it wasn't leaked there? *I* never sent them a copy. How did they get it?"

This was no time to open the wound further, so I abruptly changed the subject. "Did CBS television see you?" I asked. "We were in the press room earlier today and they wanted you for an interview."

"No, I saw no one earlier today. I'll be around tomorrow for that." Paul's responses were becoming more abrupt and distracted, as he mentally prepared himself for the fast-approaching presentation. His sense of timing was accurate, too, for Brian Maple requested that the first five speakers assume their positions at the front of the hall. We only had time to wish Paul good luck as he rose and moved to the central seat at the speaker's table.

As the distinguished panel assembled, we of the audience watched and waited for history to be made.

Chapter 2.

The Woodstock of Physics

7:30 *P.M.*, Wednesday, 18 March 1987

Though the aisles of the large hall were not yet completely cleared, Paul Chu and the other speakers in the first group were taking their places at the long front table. Meanwhile Margee and I set about finishing our two-hour-old sandwich dinners.

Everything appeared to be going smoothly. But, unbeknownst to the patiently waiting throng, a minicrisis was narrowly averted. Most of us in the audience assumed that the seating order of the first five speakers was very significant and carefully planned. That sequence, as announced in the program, appeared to follow the historical order of entry into the high-temperature superconducting game. Bednorz and Müller of IBM Zurich were clearly first, having submitted their original manuscript on La-Sr-Cu oxide in April 1986. The University of Tokyo work confirming the IBM discovery followed, with a late November submission date. No one had a problem with the first two speakers. Paul Chu was originally scheduled to speak third on the strength of his December 15 paper on the new La-Ba-Cu oxide. The Beijing and Bell Labs groups, who submitted their original findings in the second half of December, were the only other research teams to have completed studies in 1986, and so they rounded out the celebrated panel.

As a courtesy to the foreign speakers comoderator Neil Ashcroft attempted to move the Chinese speaker, Dr. Zhong-Xian Zhao, ahead of Paul. Ashcroft reasoned that the Chinese would have done the same for visitors. But it was too much of an insult to bear. Not only had Paul been ahead of the Chinese researchers every step of the way, but there was also the nagging possibility that the Beijing group had obtained some of their information through the suspected, but unconfirmed, Houston conduit. Already keyed up at the tension of the circuslike atmosphere, Chu resolved to walk out of the meeting if things were not set right. Fortunately, reason and the original programming prevailed, and virtually no one realized what had almost transpired.

The session was brought to order at 7:30 P.M. by Ashcroft, the dapper, British-accented chairman of the APS Division of Condensed Matter Physics. After welcoming the assembled throng to the "First Annual" meeting on high-temperature superconductivity, he offered a miscellany of observations on the history of superconductivity, the need for government funding, the role of the APS, and other matters that, while relevant, seemed to go on rather too long.

Ashcroft was followed by Myron Strongin, the acting editor of *Physical Review Letters*, who had been unwittingly thrust into the thick of the superconductor race. In a few short weeks eighty-nine superconductivity manuscripts had already swamped the combined *PRL* and *Physical Review* editorial offices. Strongin described the efforts to treat everyone fairly with the new, anonymous, five-member review panel that had been instituted for all superconductor papers. With the stories of Chu's alleged mistreatment already widely known, and with so many new groups active in superconductor research, *PRL* couldn't appear to be playing favorites.

The atmosphere was charged. The main event was about to begin. The evening's moderator, Brian Maple of San Diego, announced the rules like a referee at a prize fight. Each of the first five speakers had twelve minutes; a timer would signal when ten minutes were up; following the first set of talks there would be a twenty-minute question period. Though Maple's voice had the same low-key tone that I remembered from our phone conversation, his thick, black, bushy beard and bearlike physique lent authority to his words.

Alex Müller, of the pioneering Bednorz and Müller IBM team in Zurich, was first. He spoke well, with a scholarly, dignified air and a slight German accent that seemed appropriate to his understated review of the early (pre-February 1987) history of high-temperature superconductors. Only at the very end of his presentation, in the last two viewgraphs, did he allude to the most recent IBM work on the new yttrium phase. His last transparency set me on the edge of my seat. It

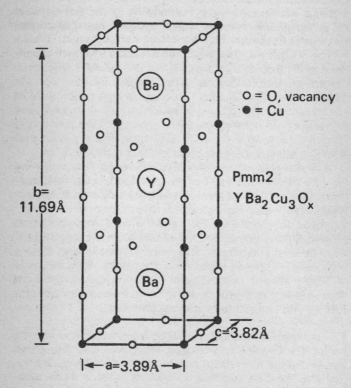

FIGURE 9. *Alex Müller closed his March 18 American Physical Society presentation by showing the IBM orthorhombic 1-2-3 structure. This structure was almost the same as ours, but their two short unit-cell edges were slightly different in length, giving a rectangular base to the structure. Our structure had a perfect square base. The oxygen occupancies were not specified in the IBM model.*

was the structure. The 1:2:3 formula and the triple-layer perovskite were clearly given; the distinctive Ba-Y-Ba ordering was there, presented as "last weekend's" results. The oxygen positions were not well defined in the illustration, but it was clear that we agreed on the basic structure.

It was also clear that we weren't the only ones who had solved it.

The University of Tokyo, represented by the distinguished Professor Tanaka, was the second speaker. In a pattern that was to become all too familiar during the long night, Tanaka began slowly with a summary of the many types of experiments done at Tokyo, but then blitzed the audience with viewgraph after viewgraph—many on display for less than ten seconds—filled with complex graphs and tables. The large Japanese group had done so much work in the past few months that there was only time to show the type of results obtained, rather than communicate the details of any one set of experiments. The frustrated audience first laughed and then groaned as the slides flashed by like license plates in the fast lane. My concentration was maintained only long enough to find out that they had misidentified the structure as the Er-Rakho type.

Paul Chu's turn came at 8:19 P.M. The audience fell silent in anticipation. Paul had much too much to say, but he used the first full minute of his precious time thanking everyone, from Bednorz and Müller for their discovery, to his dozens of colleagues and coworkers across the country, to the National Science Foundation. Art Freeman received an especially gracious acknowledgment; things were obviously patched up between them.

Paul had begun a bit nervously—hesitantly—but he quickly got into high gear with a fast-paced monologue that must have left most foreign-language speakers reeling. He unleashed a mind-boggling sequence of viewgraphs, each representing days or weeks of work, each on display for a few seconds. He soon arrived at our structure work, thanking earth scientists "who have extensive experience handling dirty samples, like rocks." Our structure illustration was flashed for only six seconds, but with the promise of a more leisurely look during my upcoming turn at the mike.

Throughout his frenetic performance Paul managed to insert whimsical asides that kept the audience off guard, and provided a good-natured counterpoint to the intense scientific experience. It was a tour de force, greeted by applause that seemed to me more intense and spontaneous than for the other speakers. Of course, I was biased.

The Chinese representative, Dr. Zhong-Xian Zhao from the Institute of Physics in Beijing, was at a double disadvantage. Not only did he have to grapple with a very foreign language, but he was also the first speaker to suffer the consequences of redundancy. With only two principal types of superconductors under discussion (the 2-1-4 compounds with the K_2NiF_4 structure and the 1-2-3 perovskites like the YBa_2Cu_3 oxides), and with relatively few different kinds of basic electrical and magnetic measurements to be made on each type, many experiments had been duplicated in many labs. The Chinese group had certainly done a lot of work, but by the fourth talk most of the viewgraphs were beginning to seem awfully familiar. Furthermore, it was obvious that the Beijing group had not yet identified the 1-2-3 composition, much less deduced its novel structure. Their data on impure, poorly characterized samples, though of tremendous significance when they were first measured two or three weeks before, now seemed obsolete. I took a mental breather, waiting for what was sure to be a blockbuster Bell Labs performance.

Bertram Batlogg did not disappoint. He razzled and dazzled and saved the best for last. He had all of the same historic electrical resistance data as the previous speakers, though on purer samples and presented on viewgraphs drafted more professionally. He also showed classy tables of more than a dozen *new* compositional variants of the 1-2-3-type phase, with superconducting transitions as high or higher than the Y-Ba original. And then he showed the structure.

The viewgraph was noticeably less professional than the previous illustrations, a sign I took to mean that the work was hot off the diffractometer. But the structure was clear enough: a triple-layer perovskite with Ba-Y-Ba ordering and obvious oxygen vacancies at the yttrium level—exactly like our structure. I quibbled with the structure drawing, which had one set of copper atoms surrounded by an octahedron of six oxygen atoms. We had found that some of those sketched-in oxygens were actually missing. What riveted my attention, however, was not the drawing but the caption: Bell Labs claimed that the structure was orthorhombic, not tetragonal. Their unit cell did not have a perfect square-shaped base, but rather an *almost*-square base with one side 1 percent longer than the other. It was a subtle difference, but one that might have profound implications for the structure and properties of the superconductor. We should not—could not—have missed such a distortion. But it was obvious that the Bell Labs crystallographers knew their business. What was going on?

I was left in an uneasy mood as Batlogg shifted from science to technology. To the astonishment of the crowd, he revealed a superconducting ceramic toroid, a 1-inch-diameter doughnut-shaped object that could form the core of a superconducting motor or electromagnet. The excited audience had barely begun to absorb the implications when he upstaged himself, unveiling a roll of superconducting foil that could be shaped and then fired to any desired configuration. Batlogg concluded simply, "I think our life has changed," as the ballroom erupted in cheers and applause.

In three short weeks the Bell Labs team had apparently bridged the gulf between scientific discovery and practical applications. It had been a virtuoso performance. While most people acknowledged that Paul Chu was the discoverer of the 1-2-3 superconducting compound, Batlogg left little doubt about who was going to be the leader, at least for a time, in the race for superconductor applications. And most of us, including Paul Chu, acknowledged that the big industrial labs would be the ones to lead the way in applying superconductivity to everyday problems.

I was inspired by the Bell Labs display, but the more I reflected on their structure drawing the more agitated I became. They, as well as the IBM and Japanese groups, seemed to be indicating octahedrally coordinated copper—six oxygen atoms surrounding copper instead of the four atoms that we had found. I had to set the record straight. As soon as Brian Maple opened the floor to questions I went to the microphone, which, being closest to the podium, was the first to be recognized.

"I wanted to comment on the structure work," I began and gave a thirty-second statement concluding that copper atoms are four-planar coordinated, not six coordinated, thus giving the structure a unique layering. There was no rebuttal from any of the speakers, and Maple, with a bemused smile and tolerant tone of voice, asked if there were any more "questions." I hoped the laughter that followed was more good-natured than derisive. In any case, I had made my point.

The questions came from every corner of the hall. How high might the temperature of superconductivity be pushed? Can a wire be fabricated? How much current will the material handle? How do you make crystals?

In the midst of this excited questioning came what was perhaps the oddest moment of the evening. Berkeley physicist Marvin Cohen read a prepared statement clarifying "inaccurate descriptions" by other speakers of Berkeley results. In the course of the short text he claimed that Berkeley, after reading of Paul's initial discovery of an "unidentified" 90-Kelvin superconductor, "independently discovered superconductivity in Y-Ba-Cu oxides." But many in the audience were not ready to accept Cohen's bold claim of independent discovery of the 1-2-3 superconductor. His statement, perhaps poorly phrased or overstated, drew a chorus of boos and catcalls, uncharacteristic of the rest of the evening's proceedings.

For much of the early presentations the rivalries between groups were only thinly veiled. Each research leader, while complimenting the others, seemed to jockey for position. Many slides carried dates in bold characters—dates that explicitly placed the moment of measurement before a rival's. But one question from the audience seemed, for a time anyway, to emphasize the kinship of the five speakers. Allen Goldman, a thin-films specialist at the University of Minnesota, asked about the failures—the compositions that didn't work. Each speaker answered in turn. Müller told of two-and-a-half years of failure before the discovery in January 1986 of the first 30-Kelvin oxide superconductor. The Tokyo spokesman told of a month of frustration synthesizing "many green stuffs," because the furnace temperature was slightly too high. Chu noted that "there are more failures than successes." But it was Batlogg from Bell Labs who said it best: "If you ask our families, they can tell you how many nights we have been away."

The animated question period finally ended after 9:30 with the meeting about thirty-five minutes behind schedule. The

main event was over and everyone needed a stretch. As the
first five speakers returned to the roped-off area and the next
five assumed their places, the large hall came alive with a din
of long-suppressed conversations and shifting chairs. A steady
stream of scientists poured out of the crowded ballroom—a
few casual observers may have heard enough, but most were
simply headed for a long-overdue rest stop after more than
four hours of standing and sitting in the mob.

During the lull a young woman dressed in a stylish carmine
outfit with black vest eased her way through the speaker section to where Margee and I were talking. I had noticed earlier
that she was the only woman in the speaker's area. Many
research teams had female members, but the speakers were
almost always the male group leaders. Only in the past decade
had significant numbers of women entered the male-domi-
nated physical sciences, and few women had yet made it to
the top.

She introduced herself with a natural smile, confident
manner, and firm handshake. "Hello, my name is Laura
Greene. I'm with Bellcore. We've done some structure work
and I'm interested in your results." Bellcore? Bell Communications Research, that was it. I wasn't sure what they did or
where their lab was located.

I rose to meet her and introduced myself. "Yes, we worked
on Paul Chu's original samples. Let me get you our flier on
the structure." I shuffled through the papers under my chair
for the illustrated sheet.

She quickly appraised the condensed version of our manuscript. "Have you submitted this structure yet?" She sounded
a bit on edge, but not as if she doubted our work.

"Yes, last week. We sent a manuscript to *PRL* on March
tenth. How about you?"

For only an instant I sensed a look of disappointment pass
over her face, but she quickly hid her emotion and merely
handed me her own preprint, "Room-temperature structure of
the 90-K bulk superconductor $YBa_2Cu_3O_{8-\delta}$ . . . received
March 13, 1987." The first author was Yvon LePage, a crystallographer in Ottawa, Canada. I flipped through the typescript, pausing at the table of atomic coordinates and the
figure of the triple-layer perovskite that they, too, had discov-

ered. So Bellcore was the fourth group, along with IBM and Bell Labs, to have deduced the composition, isolated crystals, collected x-ray data, and solved the new structure. The three-day difference in receipt dates between us and Bellcore, though clearly in our favor, seemed trivial.

But her tone of voice seemed to shift to a more aggressive level, as she asked, "So why did you work with Paul Chu, anyway?"

I was taken aback, for I interpreted her remark as a direct criticism of Paul, not to mention an indirect jab at me for being associated with him. Perhaps she just wanted to know how a Washington-based geologist ended up collaborating with a Houston-based physicist, but I couldn't suppress Freeman's warnings about the undercurrent of resentment toward Paul. And so, in my slightly paranoid state, I interpreted her question as an attack. All I could think to say was a rather inane reply: "He asked us to look at his sample. Why not?"

As Brian Maple called the next session to order she excused herself and returned to her chair, leaving me more puzzled than annoyed.

The theorists took center stage. By a neatly symmetrical coincidence there were five principal theorists working on the problem of high-temperature superconductivity; the meeting had thus naturally fallen into groupings of five speakers. Not so obvious nor symmetrical were the reasons for allotting seven minutes to each of the nonexperimentalists, as opposed to the five-minute limits on all subsequent presenters. The most frequently cited justification was that theorists can't possibly say anything in five minutes. As it was, most of them couldn't say it in seven; all but one exceeded his allotted span.

The theoretical talks were pretty incomprehensible to me and I wondered how many people in the audience really understood the subtle and not-so-subtle differences among the five. I perked up when it was Freeman's turn, however, knowing that the structure would get another showing. He began beautifully, putting the entire event into a human perspective: "I want to thank Bednorz and Müller and Paul Chu for giving me an awful lot of fun for these last three months. It's certainly been the most joyous period of my life in physics." Art had an abundance of exciting new data—much too much for a seven-minute slot. Not only had he performed numerous detailed calculations on the 2-1-4 compounds, but our structure data provided a second talk's worth of calculations on the 1-2-3 material. It was a wonderful display. At times the frustrated audience, who wanted to savor the subtleties of Freeman's findings, emitted bursts of laughter and muffled groans as a record number of images were imposed. Our hard-won structure diagram was granted a mere four seconds, hardly long enough to tell if it was right side up. But Freeman did make it abundantly clear that the layering was a key to understanding superconductivity.

Michael Schlüter, representing Bell Labs, was the last of the five speakers in the second round. His appearance was accompanied by whispered rumors of discontent at the New Jersey facility—evidently more than one Bell Labs theorist had wanted to do the honors. But if such friction existed it wasn't evident during Schlüter's presentation. Like the talk of his coworker Bertram Batlogg, it was slick and polished. He based much of his theoretical calculations on the orthorhombic structure, which was clearly illustrated in a viewgraph quite different from the one in the earlier Bell Labs talk. Two of the three copper-oxygen planes, which we observed above and below the yttrium layer, were clearly illustrated. But instead of a third copper-oxygen plane between the two barium layers, the Bell Labs figure showed a linear feature, with a chain of square-planar coppers lined up along one square edge and perpendicular to the other. That chain was consistent with orthorhombic—not tetragonal—symmetry, because the two square edges were no longer the same.

I was really worried now. Though superficially similar, our structure was different from theirs. In less than two hours I was going to have to either refute the Bell Laboratories structure or abandon ours. Neither alternative was appealing.

I ignored the theorists' question period. My mind raced to try and find a solution to the conflict, but no answer would come. Had we made a terrible blunder? Our crystals weren't all that good, but we couldn't have missed such an obvious distortion from tetragonal symmetry. Why were the two Bell Labs structure drawings different? Was Schlüter just fishing, trying to solve the structure by intuition rather than hard work?

At 10:20 P.M., now forty minutes behind schedule, the next group of five speakers filed to the front for their five-minute servings of glory. Most of the talks seemed like more of the same, but I paid special notice to the fifth offering. It was Laura Greene and the Bellcore results. The audience was perhaps beginning to tire just a bit, but their attention was riveted by the caption on one slide of the Y-Ba-Cu compound: "Material fabricated Jan. 3, 1987," weeks before Paul's first synthesis. They had synthesized the stuff but, unfortunately for Bellcore, they hadn't gotten around to measuring the superconductivity until February 25. The delay was never explained, but it demonstrated once again that virtually every experiment and idea had been duplicated by someone in those first hectic months of 1987. Greene then switched emphasis to material characterization, showing their version of the 1-2-3 structure. She was careful to state that their structure was orthorhombic, *not* tetragonal, and she showed the pattern of oxygens that was consistent with our findings. Like it or not, I was going to have to accept the fact that at least some samples of the superconductor were orthorhombic.

It was during the third brief intermission, just after the first group of five-minute papers, that an earnest young man tapped me on the shoulder. As I turned around he reached out his hand to shake mine.

"I'm Robby Beyers from IBM Almaden," he began. "Tom Shaw and I have done high-resolution electron microscopy on the yttrium phase. I have to talk to you about your structure. I think there may be some problems with it."

I had wanted to hear the upcoming talk by the Los Alamos group on new, related compositions, and Brian Maple was already calling the session to order to introduce the next suite of minilectures. But I had to see Beyers's data before my presentation. Could he really prove us wrong?

"Sure, let's go out to the hallway where we can talk," I responded. I got up and we made our way down the crowded aisle, stooping as we went to avoid blocking the large video screens.

Though I had never before met Robby Beyers I sensed something familiar in his manner and bearing. When we had made our way to the lighted corridor outside the ballroom I realized what it was. Though he was clean-shaven and obviously a few years younger than I was, I saw in him a mirror image of myself. Superficially he was of the same height, build, and hair color, but I was more struck by his similar intensity of eye contact and mode of expression. We had both been struggling with the structure problem—and we were both scared of being wrong.

Beyers had seen my flier on the structure and he began, diplomatically, by pointing out where we agreed. High-resolution electron microscopy has evolved to the point where it is sometimes possible to see features on the scale of individual atoms. Getting such images requires great skill and a superior instrument; I had long admired the scientist-artists who managed the feat. Beyers and Shaw had produced some breathtaking images. The three-layer perovskite with alternating Ba-Y-Ba sequences was confirmed beyond doubt. His photographs were also consistent with our copper positions. But there were two major points of disagreement.

11.7 Å

FIGURE 10. *IBM scientists Robert Beyers and Thomas Shaw used an electron microscope to obtain this high-resolution image of the 1-2-3 atomic layering. The regular alternation of large metal atoms, (Ba-Y-Ba)-(Ba-Y-Ba), as well as the positions of copper atoms, are easily seen in the photograph.*

First, he showed me photographs that proved his sample was orthorhombic with thousands of tiny twins. He claimed, quite rightly, that a single-crystal x-ray experiment could just average all of those tiny domains, yielding an "average" structure that was tetragonal, when in detail the structure was orthorhombic. I couldn't argue with him on that point; if the domains were smaller than a few hundred angstroms then there was no way that a single-crystal x-ray experiment could resolve the orthorhombic distortion. Though his data seemed to contradict our own findings his assertions rang a bell. There was something about his photograph of twinning, coupled with our tetragonal structure data, that was very familiar. But I didn't have time to pursue the point as Beyers was already showing me another image.

"What really bothers me is your assignment of oxygen positions," he continued. "Look at these white bands between barium atoms." He showed me a beautifully clear image in which it seemed that every metal atom could be seen. Between most of the metals were gray areas, presumably due to oxygen. Only at the copper layer between the adjacent barium atoms were there regions of white. Beyers had interpreted those white regions as evidence that *no* oxygens occur in that copper plane—a result in direct conflict with our own. Furthermore, he believed that oxygens did occupy sites in the yttrium layer, again in conflict with our report, as well as the Bell Labs and Bellcore groups. His visual evidence was certainly persuasive, but I knew that the strict interpretation of light and dark regions on such an image did not always mean atom positions. I also thought it strange that oxygen atoms, with so few electrons compared to the heavy metals, would show up so well on an electron micrograph when we couldn't see them at all in an x-ray experiment.

Not sure how to respond scientifically, I first reacted to the exquisite images that showed the unique metal layering. "These are incredible photographs! This is great work. Congratulations!"

He seemed pleased at the compliment, but wanted more. "Thanks, but what about the oxygen positions?"

"Have you calculated images to compare with the photograph?" I responded. He understood my point immediately. There is not always a one-to-one relationship between atom positions and darkened areas on an electron micrograph. Heavy atoms like barium and yttrium are obvious, but light atoms may not show up well. It is essential to calculate the expected image from a proposed structure and then compare that calculation with the actual photograph. Without such a comparison Beyers couldn't be sure, and I was willing to bet he hadn't yet had time to complete the complicated procedure.

"We're just starting to do that now," he confirmed. "We should have the results next week."

I was eager for the answer and asked him to send me a preprint as soon as possible, but given the incomplete evidence I wasn't ready to change my talk just yet. "I guess the oxygen positions are still a matter of question then," I said. "I'll be sure to mention that in my talk. Thanks for sharing your data with me."

We both hurried back to the hall to catch the ongoing talks, though I found it increasingly difficult to focus on what the other speakers had to say. I realized that my talk, as planned, wouldn't do. It was obvious that there were two aspects of our structure that were at variance with other models. First, there was the question of oxygen positions: which ones were missing? And second, what was the crystal symmetry: orthorhombic or tetragonal?

The solution came to me at 11:15, less than an hour before my talk. It wasn't an either/or proposition. Both symmetries must be correct! It suddenly occurred to me that exactly analogous behavior is shown in the common mineral feldspar, which occurs in both a higher-symmetry monoclinic form and a lower-symmetry triclinic form. The feldspar atomic structure consists of a framework of aluminum, silicon, and oxygen atoms with a sprinkling of large potassium or sodium atoms in structural cavities. If the aluminum and silicon are distributed randomly, as is the case at high temperature or in quickly cooled volcanic feldspar, then the structure can have the higher monoclinic symmetry. But if the silicon and aluminum order into a regular alternation Al-Si-Al-Si, as occurs upon slow cooling, then the structure must have the lower triclinic symmetry. In feldspars both the symmetry and the details of atom positions are controlled by cooling histories.

The exact same thing *must* be true of the superconductor.

Speaker after speaker had told of the sensitivity of superconducting temperature and transition behavior to cooling history and oxygen atmosphere. Paul Chu's superconductor samples were rapidly cooled from high temperature; those specimens thus had the higher tetragonal symmetry with oxygens distributed randomly over positions near the two adjacent barium layers. In the more slowly cooled or annealed samples of Bell Labs, IBM, and Bellcore the oxygen atoms had a chance to arrange themselves in regular planes and chains—features that reduced the symmetry of the structure from tetragonal to orthorhombic.

With the time for my talk rapidly approaching I madly set about modifying my viewgraphs, penning in additional words and symbols that subtly adjusted my interpretation to emphasize areas of agreement and qualify points of contention. I decided to abandon my carefully rehearsed speech. I was going to have to wing it.

By 11:30 it was time for the next group of speakers—*my* group of speakers. As we took our places at the front table, the others seemed as nervous as I. The speaker to my right and I conducted a futile search for water. There were plenty of cups but the pitchers were empty. Nobody was going to stop the proceedings for that; after all, we only had to speak for five minutes.

My presentation was to be the last of the five and I listened to barely a word of those first four talks. Without trying to seem too conspicuous I reshuffled my viewgraphs and continued to make notes and plan my opening sentences.

At last, just past midnight, it was my turn. I had originally planned to adopt a formal lecture style, but given the lateness of the hour, as well as my relief over the newfound resolution to the structure dilemma, a lighter touch seemed in order. I began with my bright green mineral physics T-shirt, emblazoned with the perovskite crystal structure. I hadn't the nerve to wear it, but I could use it as a visual aid.

"In the geophysics community perovskites take on a special importance," I began. "The perovskite structure for silicates is considered now to make up half the volume of the Earth." The audience gave me a reassuring laugh as I held up the bright green shirt. But it was time for business. After the obligatory credits to Paul Chu, Geophysical Lab colleagues, and the National Science Foundation, I reviewed the past month's work. The green-phase structure was flashed up for a few seconds as I wistfully noted for the record our work on the compound, "which doesn't seem to be of much interest anymore." I also showed Rus Hemley's striking single-crystal spectroscopic measurements on the new superconductor, measurements that hadn't been reported by any other group that evening. But most of my time was devoted to the black-phase crystal structure.

I began by pointing out the many structural features on which all of us—Bell Labs, Bellcore, IBM, and the Geophysical Lab—seemed to agree. We all found the triple-layer perovskite with Ba-Y-Ba ordering and coppers at the cube corners. There was less agreement on the oxygens, which were not well delimited in anyone's x-ray experiment. Three of the four groups agreed that oxygens were missing at the yttrium levels and present at the adjacent copper levels. So far so good.

The remaining points of apparent experimental difference, I proposed, were due to actual differences in the sample preparation. Our samples were cooled relatively quickly, and so we had a tetragonal form with disordered oxygens. The samples used by Bell Labs, Bellcore, and IBM were more slowly cooled and were annealed at moderate temperature. The result was orthorhombic material.

Finally, I suggested two obvious experiments—experiments that could prove my hypothesis of two different, but closely related, structures. First, I speculated that our tetragonal samples contain less oxygen than the orthorhombic samples: controlled synthesis in atmospheres having differing percentages of oxygen was needed to resolve the stabilities of each variant. Second, I proposed that the orthorhombic and tetragonal variants differ in the degree of oxygen ordering: neutron diffraction was the best way to detect such differences.

With that unrehearsed conclusion I wound up both my talk and the formal portion of the fifth session. To the accompaniment of greatly appreciated applause I resumed my position at the front table for the question period and suddenly, for the first time all day, I felt myself begin to relax. By the time I had returned to Margee and the speakers' section a few minutes later, I had shed my intense preoccupation with our structure and was ready to let my thoughts encompass the extraordinary superconductor story for a while.

Margee and I listened to several more speakers, but by 1:00 in the morning the need to stretch and to talk about the day's events was overpowering. Gathering our things, we walked as quietly and unobtrusively as we could past rows of chairs, now less than half-filled, to an exit.

Paul Chu was standing with some friends in the corridor, and we went right over to congratulate him. He looked worn out but seemed pleased by the way the session had gone. As usual, he was quick to thank the Lab for its help, and I was equally quick to thank him for the chance to do it.

"Why don't you join us for a quick snack," I suggested. "It's on us."

"Thanks, Bob, but I'm just too tired. It's been a tough day."

"But it went well, don't you think?" Margee asked.

"I'm pleased now, but who can say how it will all turn out. In any case, I'm glad its over and we can get back to work. There's still so much to do—we've only just started, really."

Paul Chu was right, of course, the superconductor problem would occupy thousands of researchers for years to come. But our job at the Geophysical Lab, for the time being anyway, was done. We could afford to relax.

For the moment, though, I felt very keyed up. I suggested to Margee that we go outside for a quick walk in front of the hotel. With all those physicists around it wouldn't be dangerous and it would help me put the day in perspective. She was happy to come along, so we made our way downstairs to the lobby and out onto Sixth Avenue.

The city was quiet and still and very beautiful. We walked in silence for a minute or so, enjoying the darkness and the cool night air. Finally, Margee asked the inevitable question. Did we win the race?

I didn't answer her immediately, but my mind spontaneously relived the last vibrant month of research. Were we first?

We may have been first by a day or two to grasp a few small fragments of knowledge. By February 27 we had isolated and identified the 1-2-3 superconducting phase. Two days later we knew of its triple-cell perovskite structure, and we had guessed its metal arrangement with the neat ordering of barium and yttrium. We refined the crystal structure of the tetragonal variant on March 8. And almost at the stroke of midnight during the epic APS superconductor session we proposed the coexistence of two distinct 1-2-3 variants—the orthorhombic *and* tetragonal phases that would be seen by many other workers in the coming weeks.

But, with so much to learn, virtually every group was the first to know something. Bellcore published the first complete orthorhombic structure of the 1-2-3 phase. IBM produced the first high-resolution electron micrographs, first demonstrated the nature of the microtwinning, and first proposed the existence of a high-temperature phase transition. And Bell Labs, with an impressive list of firsts to its credit, was the first in print with the 1-2-3 composition and tripled perovskite unit cell, as well as the first to suggest the specific type of oxygen ordering pattern that results in the novel orthorhombic copper-oxygen chains.

And even these landmark discoveries held little meaning, given the pace of the ongoing superconductor rush. Every experiment conducted in those first frantic days would be repeated with more care and more precision. All of us knew that by the time our x-ray structure results were in print they would be completely obsolete. More precise neutron diffraction experiments, systematic compositional studies, and controlled cooling measurements would have superseded all of our efforts.

To say that we "won" seems, in retrospect, much too narrow a point of view. As trite as it may sound, the real winner of the superconductor race will be mankind. Science is like an immense building, so the metaphor goes, a building built with countless blocks of information. The edifice is always growing higher and wider, and while a few blocks may be slightly larger and stick out a little higher for a time, eventually they all merge into the whole structure. The sum of science is vastly greater than the individual bits of knowledge that we scientists struggle to secure.

What the building metaphor doesn't reveal is the fun that we all have in doing our individual bits. Everyone who participated in the superconductor rush reveled in the exhilaration of the hunt. And the mammoth superconductor meeting of March 18, 1987, where all our efforts converged, was a joyous celebration—an unprecedented sharing of an explosion of knowledge in the small, once-arcane field of research that had suddenly captured the imagination of researchers all over the world.

For me, the race was over, but the superconductor adventure has only just begun.

Epilogue

The Structure

All four atomic structure models presented at the March 18 meeting—by Bell Labs, IBM, Bellcore, and the Geophysical Lab—were almost "right." All four groups independently arrived at the correct 1-2-3 composition, the correct shape of the tripled perovskite unit cell, and the correct distribution of yttrium, barium, and copper atoms. To the casual observer all of those original structure studies were indistinguishable (see Figure 11). Even so, none of the x-ray diffraction analyses could definitively position all of the oxygen atoms.

Neutron diffraction experiments provided the answers. The first neutron study was completed before the end of March by Mark Beno and colleagues at Argonne National Laboratory. They pinpointed oxygen locations and found that systematic missing oxygen atoms led to a distinctive copper-oxygen chain—a feature for which there was no experimental evidence at the time of the APS meeting. In that subtle structural detail theorists beat the experimentalists. Prior to the neutron work, Len Mattheiss, a theorist at Bell Labs, correctly guessed the distinctive chain aspect of the structure. His idea was incorporated in Michael Schlüter's structure viewgraph— the viewgraph that had caused me such concern on the night of March 18.

Within a matter of weeks the neutron diffraction work of Argonne was duplicated by researchers at half a dozen other laboratories around the world. All of the early neutron reports claimed that the structure was orthorhombic, not tetragonal as we proposed. For several weeks it appeared that our study was seriously flawed: we had missed the obvious orthorhombic distortion that everyone else had seen. More than one paper discounted our observation of a tetragonal phase.

THE STRUCTURE OF 1–2–3

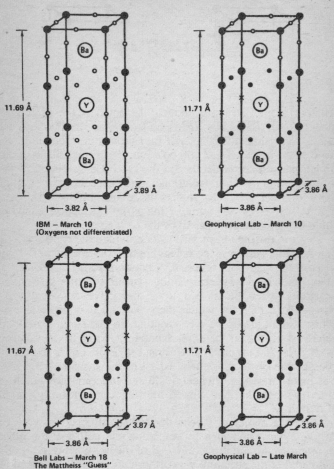

IBM – March 10
(Oxygens not differentiated)

Geophysical Lab – March 10

Bell Labs – March 18
The Mattheiss "Guess"

Geophysical Lab – Late March

FIGURE 11. *Several groups agreed on the essential features of the 1–2–3 structure. All of these models are based on the same tripled perovskite unit cell with the same arrangement of barium* (Ba), *yttrium* (Y), *and copper* ● *metal atoms. The models differ in the*

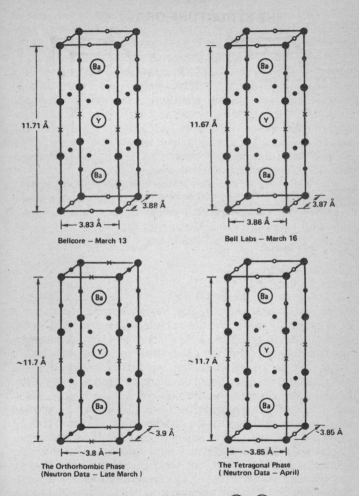

occupancies of oxygen atoms: the symbols ⊕, ○, and X indicate fully occupied, partially occupied, and unoccupied sites, respectively. The models also differ in the relative length of the two short unit-cell edges: in orthorhombic crystals the lengths differ; in tetragonal crystals they are equal.

But the structure story was not over. Even as some neutron diffraction experts were "proving" that the 1-2-3 superconductor was orthorhombic, other groups in the United States and Japan reproduced the tetragonal variety that we first observed. Robby Beyers and IBM coworkers, who suspected a possible high-temperature tetragonal variant, were able to watch the superconductor transform from the orthorhombic form to the tetragonal structure just by raising the temperature. By the end of April more than a dozen laboratories had rapidly cooled 1-2-3 to yield tetragonal crystals. Other groups showed that 1-2-3 samples poor in oxygen are almost always tetragonal, even when slowly cooled.

Gradually, after months of research and an amazing total of fifty different published 1-2-3 structure papers from a dozen countries, a logical story has emerged: there are two main structural varieties of 1-2-3, varieties that differ only by the number and location of oxygen atoms.

The Publication

We had about two weeks following the APS meeting to make minor revisions to our *Physical Review Letters* paper on the black-phase structure. We didn't collect new data because the crystals weren't going to get any better. But we did reanalyze our data in light of the Bell Labs, IBM, and Bellcore results.

Our biggest concern was the apparent confusion over the two structural variants. Were our samples tetragonal or orthorhombic? We subjected all our unit-cell data to intense analysis. Three of our original five black-phase crystals showed absolutely no evidence for the orthorhombic distortion seen by the other labs; we had to report those crystals as tetragonal. The two smallest crystals in Paul's second batch, however, revealed a tiny distortion: one short axis was ever so slightly longer than the other. The 0.15 percent difference—a variation *almost* insignificant compared to our experimental errors —was much smaller than the 1 percent or greater difference in axial lengths reported by the other groups. We probably wouldn't have bothered to report the observation under ordinary circumstances. But in light of the tetragonal-orthorhombic debate even this small amount of distortion was definitely worth noting. We added the revised axial dimensions to the

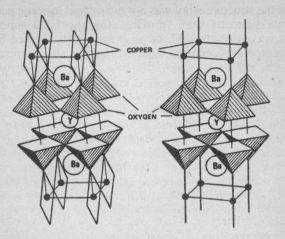

FIGURE 12. *Stylized polyhedral drawings of the orthorhombic (left) and tetragonal (right) variants of the 1-2-3 structure emphasize the subtle differences in oxygen distributions.*

manuscript and stated: "Batch 2 crystals deviate slightly from tetragonal dimensionality." Even with this acknowledgment of a possible subtle distortion, our results were at variance with those of Bell Labs, IBM, and Bellcore.

We also puzzled over the occupancies of oxygen atoms near barium atoms. In our first structure drawing we had decided to emphasize the copper-oxygen layered aspect of the structure by highlighting copper-oxygen bonds in planes. Additional partially filled oxygen sites were also designated, but with "X"'s rather than "O"'s. Reexamination of the data, however, convinced us that those oxygens designated X were actually more fully occupied than the oxygens in the copper plane. Larry quickly created a new structure figure with shaded and open circles representing fully and partially occupied oxygens, respectively. The revised manuscript was mailed to the *PRL* offices in early April.

We felt that our structure study would be published quickly in *Physical Review Letters*. After all, we had identified the unique superconductor structure. But the journal editors were in a quandary, for they had received several structure papers in a seven-day period just before the March 18 meeting. All five couldn't appear in *PRL*—there just wasn't room. And the editors didn't want to select only one for special treatment. So all five papers were bumped to the "Rapid Communications" section of the much more bulky (and less prestigious) *Physical Review B*. All five appeared together in the May 1 issue. At the time we were unhappy with the decision, feeling that our paper was the first to give crystallographic descriptions of both phases in the historic Houston/Alabama sample. Our work deserved a better fate. While our arguments did not succeed, we were able to negotiate for a "Note added," which gave us the chance to make a few last-minute additions.

In a neatly parallel decision, our hydrogen paper was also demoted from *PRL* to *Physical Review B*, where it appeared as a "Brief Report" in the September 1 issue.

The Scientists

The superconductor race continues unabated. Thousands of scientists and engineers have spent most of their time since March 1987 in the quest. By even the most conservative calculations millions of hours of scientific research and engineering have been devoted to the understanding and exploitation of 1-2-3. Hundreds of new scientific papers are submitted every month in what has become perhaps the most redundantly studied material in history.

For many researchers high T_c remains an all-consuming, seven-day-a-week passion. Art Freeman, Paul Chu, and their colleagues maintain an exhausting schedule of research, conferences, and lectures. Dozens of scientists at Bell Labs, IBM, Argonne, Bellcore, and a host of other laboratories have committed themselves to the quest. They know there is much left to do. The complexities of creating commercially viable Y-Ba-Cu oxide superconductors are becoming clear. Samples must be synthesized and handled with great care to obtain the optimum oxygen content and the best-ordered arrangement of oxygen atoms in the crystal structure. High-temperature superconductors can be cooked up in high school chemistry class, but it will take sophisticated manufacturing techniques to ensure quality products.

Superconductivity in 1-2-3 still seems to baffle the theorists. According to one widely repeated joke, it is a problem with far more theories than theorists, because everyone is trying to lay claim to multiple possibilities. But while scientists struggle to understand and increase high T_c, engineers are well on the way to practical applications. Several firms are marketing 1-2-3 superconducting disks for educational laboratory demonstrations and the Y-Ba-Cu oxide is also commercially available in powdered form. A number of companies claim to have designed superconducting wire, and one Massachusetts-based firm plans to be selling wire by the end of 1988. Perhaps of greatest significance are advances in producing high-T_c superconducting thin films suitable for applications in computers.

International competition is intense. It is said that by mid-January 1988, Japanese companies had filed more than 2000 patents and had spent more than $30 million on studying the new superconductors. As the 1-2-3 story shifts from scientific research to commercial development it is clear that a small army of patent lawyers will be among the chief beneficiaries.

The Nobel Prize

On October 14, 1987, the Nobel Prize in Physics was awarded to Karl Alex Müller and Johann Georg Bednorz for their discoveries in high-temperature superconductivity. The Royal Swedish Academy gave science's highest honor for a discovery that had been virtually unknown twelve months earlier. The unprecedented speed with which the Nobel Prize was conferred is just another example of the unique pace of the ongoing superconductor saga. In spite of the obvious Nobel caliber of the seminal Bednorz and Müller work, it was an announcement that caught many physicists by surprise. By awarding the prize in 1987 the Royal Swedish Academy effectively blocked any possible consideration of Paul Chu, whose discovery was announced after the January 31 nomination deadline.

The selection of Bednorz and Müller was appropriate and it was clever. Everyone acknowledges the IBM researchers as the first to observe superconductivity in the copper-bearing oxide compounds. There is no controversy there. The Nobel citation justly credited the two IBM physicists for inspiring "a great number of other scientists to work with related materials." Many of us perceive that phrase as a particular reference to Paul Chu, for without the discovery of 90-K superconductors it seems unlikely that the Nobel committee would have acted so quickly to honor a 30-K superconductor. The nomination deadline of January 31 was apparently only barely met by the two IBM Zurich scientists. All subsequent superconductor advances, including Chu's discovery, were not announced until after February 1, and they were thus not eligible for the 1987 award. By giving the 1987 prize to Bednorz and Müller the Nobel committee recognized the profound importance of high-T_c superconductors, but managed to avoid the potential controversy over who did what in the later mad rush.

Paul Chu thus did not *lose* the Nobel—he wasn't even a contender. And Chu is by no means out of the running for a Nobel Prize. His record-breaking achievement of a practical, liquid-nitrogen–temperature superconductor is the key discovery that sent thousands of scientists and engineers scurrying to their labs. As superconductors begin to affect our lives it seems likely that the 1-2-3 discovery will be recognized as the breakthrough. And so, each October, superconductor researchers in Houston will be hoping for a phone call from Stockholm to Houston.

The Future

We seem to be poised on the verge of a new technological age. Superconductors are not likely to become an obvious presence in our society, anymore than electric motors or microchips are obvious in day-to-day living. We take these marvels for granted. But every home has dozens of electrical motors—in razors and refrigerators, toys and toothbrushes. We use microchips in everything from televisions and telephones to microwaves and thermostats. In a similar way superconductors will gradually insinuate themselves into our lives.

Some applications will be obvious extensions of well-known technologies, technologies that are too expensive to improve with low-temperature superconductors. Superconducting motors could be smaller and more powerful than conventional designs. Battery-powered automobiles with light, efficient superconducting electric motors could revolutionize society and its dependence on fossil fuels. Superconducting computers will be smaller and faster. We could have desktop home computers with computational and graphics capabilities as sophisticated as current room-size models. Such small, fast computers might lead, for example, to new generations of interactive video games with stunning three-dimensional visual displays and intricacies of play vastly more sophisticated than current models. Superconducting communications lines could carry a hundred times the data of optical fibers: imagine color video telephones with crystal clarity, or the near-instantaneous transmission of the latest best-seller from publisher to home printers. With superconductors we will be able to do more things—and do them faster and cheaper—than ever before.

Much of the ongoing 1-2-3 research is an effort to increase the "critical current," the amount of electrical current above which superconductivity ceases. The critical current of most 1-2-3 samples is too small for practical applications in power generation and transmission. If the critical current problem is resolved then superconductor power plants, cables, and storage coils could revolutionize the production, distribution, and storage of electrical power.

Superconducting magnets, already the key to a number of expensive technologies, would be much cheaper to operate with high-T_c materials. One major market for these magnets is in the production of medical scanning devices (MR—magnetic resonance—scanners) that allow physicians to obtain a detailed three-dimensional image of the human body without x-rays. Cheaper and faster medical scanners would make this diagnostic procedure widely available and thus could help save lives. The same superconductor magic, if manipulated to the requirements of space-based laser weaponry, could have the opposite effect. In what sounds like a futuristic nightmare, but is a prospect already on the drawing boards, engineers foresee the development of superconductor infrared sensors so accurate that they will be able to detect—and target—individual humans on earth from space. For legions of scientists, engineers, and policy planners, the perfection of such superconducting technologies and the judicious integration of these devices into our society will provide a focus for their professional lives for years to come.

Woodstock II

For almost a year after Chu's discovery scientists wondered if it was possible to achieve T_c even greater than 90 K. Unconfirmed reports of new compounds with superconductivity at higher onset temperatures abounded during the spring and summer of 1987. A Michigan company reported 155 K for a variety of 1-2-3 in which fluorine replaces some of the oxygen. A North Carolina research team claimed to have isolated a 240-K material. And dozens of researchers were tantalized by transient effects suggestive of superconducting temperatures up to—and even above—room temperature! But none of these observations was easily reproducible. Thousands of researchers spent months on the problem while Paul Chu held the high-T_c record for a stable superconducting compound.

In late January 1988 Paul Chu broke his own record when he discovered a stable, reproducible 120-K superconductor. Chu's team, intrigued by reports from France of 22-K superconductivity in an oxide of bismuth, strontium, and copper, began to study chemical combinations of bismuth and copper. Shortly thereafter, laboratories in Houston and Tsukuba independently found superconductivity in a new oxide chemical system containing bismuth, calcium, strontium, and copper. That mixture, they soon learned, yields a superconductor completely different from 1-2-3. A new race was on to characterize the unknown bismuth-bearing superconductor and to optimize its behavior.

Those of us at the Geophysical Laboratory who were involved in the 1-2-3 frenzy followed the weekly superconductor updates with an avid, almost proprietary interest. But, as far as our own experiments were concerned, it was back to business as usual as we settled into the comfortable, less frenetic study of rocks and minerals (especially perovskites). We thought that an exciting, but thoroughly atypical, episode was over.

We were wrong. The chaos began again on Tuesday, January 26, 1988, when Paul Chu's newest superconducting samples were expressed to us. A few weeks later yet another new superconductor, this one laced with the element thallium, arrived from Professor Allen Hermann at the University of Arkansas. For a frantic month we lost sleep over bismuth and thallium instead of yttrium, calcium instead of barium. The cycle began again as laboratories around the world focused on the latest discovery of superconductivity above 100 K. In February 1988 the Geophysical Laboratory team discovered three new superconductors—all with T_c above 100 K—and this time we didn't wait to tell the world. Our announcements were followed within days by confirmation and further discoveries at a dozen laboratories.

The new superconductors took center stage on March 22, 1988, at the annual meeting of the American Physical Society, held this time in New Orleans. The special last-minute, marathon session was dubbed "Woodstock II," as dozens of researchers from around the world shared their latest results with a standing-room only crowd of more than 700 scientists and reporters. It was the end of another race, another "atypical" episode. But how atypical was it? Certainly the social significance of Paul Chu's historic compounds, as well as the extraordinary acceleration of the scientific process during the early months of 1987 and 1988, sets the Geophysical Lab's superconductor experiments apart from our usual research routine. On the other hand, the quest for the structures of the superconductors could be described as just another job. Someone poses a question and we, with our experimental hardware and computer programs, have to figure out an answer. We do it every day. Although we usually work on natural samples of earth-bound materials, the procedures used to unravel the mysteries of superconductors are standard. So is the exhilaration at finding a solution.

Part of the excitement of science has always been the satisfaction that comes when an answer, however small, emerges from the data. The other part is that the questions never stop coming.

Acknowledgments

During the writing of this book I have received generous assistance from many principal players. It would be impossible to credit all of the dozens of researchers who have helped me write this book; they all have my gratitude.

Paul Chu, in particular, has taken hours of his time to detail the historic events leading up to the discovery of the 1-2-3 phase, as well as the extraordinary atmosphere in the two months prior to the March 18 meeting of the American Physical Society. He has been candid about both his successes and failures, and he has consistently taken more pleasure in praising his colleagues than talking about himself. Paul's skill in coordinating his transcontinental, multisite research team is ample evidence of his gracious good nature and scientific leadership.

Paul Chu's Houston colleagues, including Jeffrey Bechtold, Ken Forster, Li Gao, Pei-Herng Hor, Zhi-Jun Huang, Theresa Lambert, Ru-Ling Meng, Simon Moss, and Ya-Qin Wang, graciously shared their knowledge of the hardware and history of the Houston laboratory. Eric Miller of the University of Houston's Public Relations Office supplied background information on the history of the campus and he provided newspaper reports and photographs. Special thanks are due May Chu, who not only reviewed the manuscript, but also proved a gracious hostess as I monopolized Paul during one of his infrequent evenings at home in Houston.

Professor Arthur Freeman, who enjoyed the excitement of Paul Chu's discovery almost from the beginning, has provided invaluable insights and opinions on the developments of late 1986 and early 1987. As Paul Chu's confidant, Art was in the unenviable position of being forced to withhold crucial information from his Northwestern departmental colleagues. Yet, in spite of the misunderstanding that almost severed his contact with Houston, Freeman never betrayed Chu's trust. A love for physics was always his primary motivation; it is a pleasure to call Art my friend.

I am indebted to the Carnegie Institution of Washington for its continued help and encouragement. My coworkers at the Geophysical Laboratory—Ross Angel, Larry Finger, Rus Hemley, Dave Mao, Charlie Prewitt, and Nancy Ross—have given me their constant support and friendship. Each reviewed the book manuscript with meticulous care for historical accuracy and an open-minded sympathy for my interpretation of the facts. Ray Bowers and Patricia Parratt of the Carnegie Institution's Editorial Office also contributed careful critiques of the penultimate draft.

I have received thoughtful and constructive reviews of the manuscript from a number of key scientists in the superconductor story. Maw-Kuen Wu, professor of physics at the University of Alabama, detailed the events surrounding the first synthesis of the 1-2-3 compound and explained aspects of the cooperation between Houston and Huntsville.

Myron Strongin, acting editor of *Physical Review Letters* during the first months of the superconductor rush, clarified the central role of *PRL* during the period from December 1986 through March 1987. Strongin and his colleagues Peter Adams, David Lazarus, Kelvin Lynn, Per Bak and Reid Terwilliger provided forthright assessments of the intense competition and resultant controversies surrounding the flood of manuscripts submitted during the first months of the superconductor story.

Dr. Robert Dynes, who played a prominent role in the extraordinary mobilization of AT&T Bell Laboratories following the discovery of high T_c, provided valuable details on the chronology of discoveries at the New Jersey facility. He, along with Bertram Batlogg and Robert Cava, improved the manuscript with a thoughtful commentary based on their unique view of the winter of 1986–87.

Paul Grant, Robert Beyers, and Edward Engler of IBM's West Coast Almaden Laboratory shared their personal knowledge of the Bednorz and Müller story, as well as subsequent IBM activities on the 1-2-3 compounds. Through candid and constructive reviews of the manuscript, each of these scientists was able to communicate a perspective of the superconductor race different from my own.

It was a special privilege to receive a manuscript review from Professor Philip Anderson, theoretical physicist at Princeton University. Though initially skeptical of my approach, Anderson reviewed the text with an open mind and contributed a constructive and supportive analysis.

Other individuals who reviewed portions of the book include Laura Greene of Bell Communications Research, and David Nelson and Robert Pohanka of the Office of Naval Research. They, too, have my thanks.

I am indebted to a number of nonscientist reviewers, including Dr. Gary Taylor, professor of English at The Catholic University of America, and my parents, Peggy and Dan Hazen, who helped to steer me clear of opaque scientific jargon; too often in science we employ a jumble of incomprehensible (and sometimes meaningless) words.

I owe a great debt to Helen Morris Hindle for helping to define the concept of this book and to both Helen and Brooke Hindle for encouraging and advising me along the way.

Much of the book was written during a summer sabbatical at the chemistry department of the University of California at Santa Barbara. My gracious host, Professor Galen Stucky, who welcomed me as a friend and colleague and provided a splendid environment to think and write, also contributed a valuable critique of an early draft of the book.

The concept and format of the book was developed jointly with my agent, Gabriele Pantucci, who has been a continuous source of encouragement. James Silberman, of Summit Books, supported the project enthusiastically and helped define its scope and style. The manuscript was polished with craftsmanship and patience by my editor, Dominick Anfuso. His artful combination of tact and candor are deeply appreciated.

My principal debt is to my wife, Margaret Hindle Hazen, sometime coauthor and always best friend. She contributed substantially to every draft of the manuscript, criticizing and complementing, editing and revising. The entire history of this book, from conception to outline to writing, bears her imprint along with my own.

Afterword to 1989 Edition

Gradually, as the world of materials research returns to a saner state, a picture of the true superconductivity winners and losers begins to emerge. A few scientists have gained a justifiable measure of fame. Many more will get rich, though the monetary rewards may have little to do with scientific merit.

George Bednorz and Karl Alex Müller were the first and most obvious winners. The 1987 Nobel Prize in Physics is theirs, thanks to their discovery which triggered the whole superconductor race. Though widely applauded, their Nobel seems in retrospect an unusual award. True, the IBM discovery was in the realm of superconductivity, traditionally a part of physics. But Bednorz and Müller's advance was the development of a new chemical compound—previously the exclusive domain of chemistry. Never before had a physics Nobel Prize been awarded for a new material (though it may be argued that never before have so many physicists suddenly taken up materials research). The Bednorz and Müller Nobel Prize dramatizes the arbitrary nature of traditional boundaries in the physical sciences.

Thousands of patents have been filed for high-temperature superconductors and their applications, but no one seems very concerned about patent priorities in the case of the Bednorz and Müller superconductor. Their record breaking 2-1-4 material has been eclipsed and engineers entertain no prospects for practical applications there. Few will disagree that Bednorz and Müller were the only discoverers of superconducting 2-1-4.

The 1-2-3 story is not so simple. Paul Chu, as leader of the Houston/Alabama effort that produced the first liquid-nitrogen-temperature superconductor, has received the most attention. Last year Chu was awarded a Presidential Science Medal, as well as prizes by the American Institute of Physics and the New York Academy of Sciences. Some researchers counter, however, that Alabama physicist Maw-Kuen Wu, or perhaps even his graduate students, deserve a greater share of the glory. After all, they did do the critical hands-on grinding and baking of chemicals that produced the first magic 1-2-3 samples. Why should Chu get credit for work at another laboratory?

The answer is simple and reflects longstanding traditions in science. Paul Chu was the acknowledged team leader, and the team leader gets top billing. Chu invited Wu to help in the search. Chu outlined a research strategy that included substituting yttrium for lanthanum. Paul Chu is entitled to a major share of the credit.

Whatever the outcome of the unfortunate Houston/Alabama tussle, 1-2-3 scientific prizes seem small potatoes compared to the potential patent rights. Already Dupont has given the University of Houston $1.5 million for licensing rights to 1-2-3, with the promise of millions more should Houston be granted the patent. But in spite of the obvious central role of Houston and Alabama physicists, the ultimate winners in the 1-2-3 patent case are not at all obvious.

An officer of the United States patent courts recently ruled that the 1-2-3 rights will be given to the first group to have demonstrated five things: (1) high-temperature superconductivity; (2) a yttrium-barium-copper-oxygen material; (3) the 1-2-3 composition; (4) the perovskite atomic structure; and (5) at least 90% pure form. Many of us are puzzled by this policy. The first two criteria alone are the essence of Chu's discovery. High-temperature superconductivity in the yttrium mix is all you really have to know. That was the subject of the Houston/Alabama paper in *Physical Review Letters*, and those details were made public by Chu in his Santa Barbara lecture of 26 February, 1987. The composition and characteristics of 1-2-3 were given freely to dozens of laboratories around the world. Armed with the yttrium recipe, any competent materials research lab could apply routine procedures to synthesize and analyze material (criteria 3 to 5). But with the patent office's new, demonstrably unfair ruling, giants like IBM and Bell

Labs have staked their claim, saying they were first to synthesize the 90% pure stuff. The big boys, in an effort to push Chu and Wu aside, say they were the ones to "invent" 1-2-3.

Larry and Rus and Charlie and I thought our contribution to the 1-2-3 saga was over two years ago, but now we're caught in the middle of a patent fight. (The Rosses, now married, have both landed jobs at University College, London, and so have escaped the legal complications.) Working with Paul Chu, the Geophysical Laboratory was the first to isolate crystals of pure 1-2-3, to measure its composition, and determine its perovskite-type structure. Our work may thus be a keystone in the Houston patent case. So now enter the lawyers.

In January of 1989 a team of three conservatively suited, middle-aged men from the University of Houston and Dupont marched into my office with copies of my book clutched in their hands. These weren't literary types hoping for an autograph. To them, *The Breakthrough* is a preliminary legal brief, and every scrap of computer output, lab notebook, or structure doodling in our files is a legal document, to be stamped with Bates numbers and scrutinized by an army of aides. Box upon box of snippits—newspaper clippings, meeting abstracts, preprints, reprints, and even my book reviews —have become ammo for the upcoming legal battle.

It may be years before the case is settled, but even the winner of the 1-2-3 fight may not come away with much. Two challenges confront the promoters of 1-2-3. First, other potentially better superconductor recipes have come along. In January of 1988 Allen Hermann, professor of physics at the University of Arkansas, sent us a small chip of a new black superconductor. This time the magic element was thallium (mixed with more familiar things like barium and copper and oxygen). And this time the temperature was above 120 Kelvin, shattering the year-old mark by more than 20 degrees.

It was déja vu. Another fine-grained black sample. Another mad race to measure composition and structure. Within a week of receipt of those samples we had the answers, neatly typed, submitted for publication. We immediately telefaxed the manuscript to more than a dozen laboratories around the world to stifle any conflicting claims of priority. Robby

Beyers and Paul Grant at IBM's Almaden, where other thallium compounds had been studied for years, were able to duplicate the Arkansas results in less than twenty-four hours and announced the results at a major superconductivity conference at Interlachen, Switzerland, the next day. This time there can be no question of who wins. Hermann, Sheng, and the University of Arkansas expect preliminary licensing fees approaching $5 million.

An even more daunting roadblock stands in the way of a 1-2-3 windfall. Despite the extraordinary potential of these new materials, the initial euphoria of high-temperature superconductivity has largely disappeared. 1-2-3 and its cousins have fabulous properties, but they are the devil to manufacture into useful forms. Unlike metals, which can be melted, molded, or pulled into wire, the new ceramic superconductors are brittle like procelain. Entirely new processing procedures must be developed, so it will be years, perhaps decades, before they begin to insinuate themselves into our daily lives. Applications in microcircuitry, magnetic levitation, and in space (where no special cooling is required) may come by 1995. But the big-ticket items—magnetic trains, power storage rings, superconducting cables, and superefficient motors —may not be seen until the next millenium. In fact, the seventeen-year limits on superconductor patents may well have expired before any widespread applications are on line.

Who are ultimate winners of the high-temperature superconductivity race? In the short term a few scientists and engineeers, a couple of universities, a bunch of patent lawyers, and perhaps an author or two will profit. But the real winners, down the road a ways, are children of the next generation, for whom the phenomenon of a levitating magnet will be as commonplace as lasers and personal computers. For them, high-temperature superconductors will be one more useful material with which to shape a world. Superconductors will allow them to do more things, faster and safer, than before. It may take time, but given these remarkable, revolutionary new materials, time is all that it will take.

Robert M. Hazen, February, 1989

INDEX

About the Author

Robert Hazen, a graduate of MIT and Harvard, is a research scientist at the Carnegie Institution of Washington's Geophysical Laboratory, where he and his colleagues study the structure of minerals and their properties. He has written books and articles on a variety of subjects, including crystal chemistry and solid-state physics, geological poetry, the history of mining, and American brass bands. A part-time professional trumpeter, Hazen has performed with a number of ensembles including the National Symphony, the Metropolitan Opera, and the Royal Ballet. He lives in Bethesda, Maryland, with his wife, author Margaret Hindle Hazen, and their two children, Benjamin and Elizabeth.

FACT

is stranger than

FICTION